大数据技术丛书

Spark
入门与大数据分析实战

迟殿委 李超 著

清华大学出版社
北京

内 容 简 介

本书基于 Spark 3.3.1 框架展开，系统介绍 Spark 生态系统各组件的操作，以及相应的大数据分析方法。本书各章节均提供丰富的示例及其详细的操作步骤，并配套示例源码、PPT 课件和教学大纲。

本书共分 11 章，内容包括 Scala 编程基础、Spark 框架全生态体验、Spark RDD、Spark SQL、Kafka、Spark Streaming、Spark ML、Spark GraphX、Redis 等技术框架和应用，并通过广告点击实时大数据分析和电影影评大数据分析两个综合项目进行实战提升。

本书适合 Spark 框架初学者，既可以作为大数据分析技术、大数据应用开发工程师的查询手册，也可以作为高等院校或高职高专计算机技术、软件工程、数据科学与大数据科学、智能科学与技术、人工智能等专业大数据课程的教材。

本书封面贴有清华大学出版社防伪标签，无标签者不得销售。
版权所有，侵权必究。举报：010-62782989，beiqinquan@tup.tsinghua.edu.cn。

图书在版编目（CIP）数据

Spark 入门与大数据分析实战 / 迟殿委，李超著. —北京：清华大学出版社，2023.6
（大数据技术丛书）
ISBN 978-7-302-63798-1

Ⅰ. ①S… Ⅱ. ①迟… ②李… Ⅲ. ①数据处理软件—教材 Ⅳ. ①TP274

中国国家版本馆 CIP 数据核字（2023）第 105797 号

责任编辑：夏毓彦
封面设计：王　翔
责任校对：闫秀华
责任印制：沈　露

出版发行：清华大学出版社
　　　　　网　　址：http://www.tup.com.cn，http://www.wqbook.com
　　　　　地　　址：北京清华大学学研大厦 A 座　　邮　编：100084
　　　　　社 总 机：010-83470000　　　　　　　　邮　购：010-62786544
　　　　　投稿与读者服务：010-62776969，c-service@tup.tsinghua.edu.cn
　　　　　质 量 反 馈：010-62772015，zhiliang@tup.tsinghua.edu.cn

印 装 者：三河市科茂嘉荣印务有限公司
经　　销：全国新华书店
开　　本：190mm×260mm　　　印　张：16.25　　　字　数：438 千字
版　　次：2023 年 7 月第 1 版　　　　　　　　　　印　次：2023 年 7 月第 1 次印刷
定　　价：79.00 元

产品编号：102832-01

前　　言

如今大数据技术已广泛应用于金融、医疗、教育、电信、政府等领域，各个行业都积累了大量的历史数据，并不断产生大量新数据，数据计量单位出现 PB、EB、ZB、YB，甚至 BB、NB、DB。大数据的处理方式与传统数据不同，需要通过分布式存储和分布式运算来实现，由此也催生了优秀的大数据处理框架和生态组件。Spark 的特色在于它首先为大数据应用提供了一个统一的平台。从数据处理层面看，模型可以分为批处理、交互式、流处理等多种方式；而从大数据平台层面看，已有成熟的 Hadoop、Cassandra、Mesos 以及其他云的供应商。Spark 整合了主要的数据处理模型，并能够很好地与现在主流的大数据平台集成。

许多大型互联网公司，如谷歌、阿里巴巴、百度、京东等都急需掌握大数据技术的人才，因此大数据相关人才出现了供不应求的状况。Spark 作为继 Hadoop 之后的下一代大数据处理引擎，经过飞跃式发展，现已成为大数据产业中的一股中坚力量：RDD 模型具有强大的表现能力，并通过不断完善自己的功能而逐渐形成了一套自己的生物圈，提供了全栈（full-stack）的解决方案，其中主要包括 Spark 内存中批处理、Spark SQL 交互式查询、Spark Streaming 流式计算、GraphX 图计算和 Spark ML 机器学习算法库。

关于本书

本书基于 Spark 3.3.1 新版本展开，符合企业目前的开发需要。本书全面讲解 Spark 大数据技术的相关知识和实战应用，内容包括 Scala 编程基础、Spark 框架全生态体验、Spark RDD、Spark SQL、Spark Streaming、Kafka、Spark GraphX、Spark ML、Redis 等技术框架及其应用，并通过广告点击实时分析和电影影评分析两个大数据分析综合项目进行实战提升，夯实 Spark 大数据分析的基础知识，提升开发技能。

本书重视实践操作开发，内容安排从框架搭建和开发环境安装、技术框架快速示例引入、技术框架详细案例讲解，到大数据分析综合项目实战提升等，将实战与理论知识相结合，从而加深读者对 Spark 框架应用的理解。

笔者是具有多年大数据分析和处理实战经验的高级工程师，书中融入了笔者多年的实战经验，讲解细致、内容丰富、示例清晰、语言通俗易懂，方便读者提高学习效率，保证学习质量。

配套示例源码、PPT 课件等资源下载

本书配套示例源码、PPT 课件、教学大纲，需要用微信扫描下边二维码获取。如果下载有问题或阅读中发现问题，请用电子邮件联系 booksaga@163.com，邮件主题为"Spark 入门与大数据分析实战"。

适合的读者

- Spark 框架初学者。
- 大数据分析技术人员。
- 大数据应用开发工程师。
- 高等院校或高职高专大数据课程的师生。

笔　者

2023 年 3 月

目 录

第1章 Spark 开发之 Scala 编程基础 ············· 1
- 1.1 开发环境搭建 ············· 1
- 1.2 基础语法 ············· 4
- 1.3 函数 ············· 7
- 1.4 控制语句 ············· 9
- 1.5 函数式编程 ············· 12
- 1.6 模式匹配 ············· 17
- 1.7 类和对象 ············· 18
- 1.8 异常处理 ············· 22
- 1.9 Trait（特征） ············· 23
- 1.10 文件 I/O ············· 24

第2章 Spark 框架全生态体验 ············· 26
- 2.1 Spark 概述 ············· 26
  - 2.1.1 关于 Spark ············· 26
  - 2.1.2 Spark 的基本概念 ············· 27
  - 2.1.3 Spark 集群模式 ············· 28
- 2.2 Linux 环境搭建 ············· 33
  - 2.2.1 VirtualBox 虚拟机安装 ············· 33
  - 2.2.2 安装 Linux 操作系统 ············· 35
  - 2.2.3 SSH 工具与使用 ············· 42
  - 2.2.4 Linux 统一设置 ············· 43
- 2.3 Hadoop 安装与配置 ············· 45
  - 2.3.1 Hadoop 安装环境准备 ············· 45
  - 2.3.2 Hadoop 伪分布式安装 ············· 49
  - 2.3.3 Hadoop 完全分布式环境搭建 ············· 55
- 2.4 Spark 安装与配置 ············· 60
  - 2.4.1 本地模式安装 ············· 61
  - 2.4.2 伪分布模式安装 ············· 63
  - 2.4.3 完全分布模式安装 ············· 66
  - 2.4.4 Spark on YARN ············· 68
- 2.5 spark-submit ············· 72
  - 2.5.1 使用 spark-submit 提交 ············· 72
  - 2.5.2 spark-submit 参数说明 ············· 73
- 2.6 DataFrame ············· 75
  - 2.6.1 DataFrame 概述 ············· 75
  - 2.6.2 DataFrame 的基础应用 ············· 77

2.7 Spark SQL ··· 82
   2.7.1 快速示例 ··· 83
   2.7.2 read 和 write ··· 87
2.8 Spark Streaming ·· 89
2.9 共享变量 ·· 92
   2.9.1 广播变量 ··· 92
   2.9.2 累加器 ·· 93

第 3 章 Spark RDD 弹性分布式数据集 ··· 94
3.1 什么是 RDD ··· 94
3.2 RDD 的主要属性 ··· 95
3.3 RDD 的特点 ··· 96
   3.3.1 弹性 ··· 96
   3.3.2 分区 ··· 96
   3.3.3 只读 ··· 96
   3.3.4 依赖（血缘）··· 96
   3.3.5 缓存 ··· 98
   3.3.6 checkpoint ·· 99
3.4 RDD 的创建与处理过程 ·· 99
   3.4.1 RDD 的创建 ··· 99
   3.4.2 RDD 的处理过程 ·· 99
   3.4.3 RDD 的算子 ··· 100
   3.4.4 常见的转换算子 ··· 100
   3.4.5 常见的行动算子 ··· 105

第 4 章 Spark SQL 结构化数据文件处理 ·· 109
4.1 Spark SQL 概述 ··· 109
   4.1.1 什么是 Spark SQL ·· 109
   4.1.2 Spark SQL 的特点 ·· 110
   4.1.3 什么是 DataFrame ·· 111
   4.1.4 什么是 DataSet ·· 112
4.2 Spark SQL 编程 ··· 112
   4.2.1 SparkSession ·· 112
   4.2.2 使用 DataFrame 进行编程 ·· 113
   4.2.3 使用 DataSet 进行编程 ·· 118
   4.2.4 DataFrame 和 DataSet 之间的交互 ··· 120
   4.2.5 使用 IDEA 创建 Spark SQL 程序 ··· 120
   4.2.6 自定义 Spark SQL 函数 ··· 121
4.3 Spark SQL 数据源 ·· 122
   4.3.1 通用加载和保存函数 ··· 122
   4.3.2 加载 JSON 文件 ··· 123
   4.3.3 读取 Parquet 文件 ·· 124
   4.3.4 JDBC ·· 124

第 5 章 Kafka 实战 ··· 127
5.1 Kafka 的特点 ·· 128
5.2 Kafka 术语 ··· 129
5.3 Kafka 单机部署 ·· 130
5.4 Kafka 集群部署 ·· 137

第 6 章 Spark Streaming 实时计算 ··· 142
6.1 Spark Streaming 概述 ·· 142
6.1.1 Spark Streaming 是什么 ·· 142
6.1.2 Spark Streaming 特点 ··· 143
6.1.3 Spark Streaming 架构 ··· 144
6.2 DStream 入门 ·· 144
6.2.1 WordCount 案例 ·· 145
6.2.2 WordCount 案例解析 ··· 146
6.3 DStream 创建 ·· 147
6.3.1 RDD 队列 ·· 147
6.3.2 自定义数据源 ·· 148
6.3.3 Kafka 数据源 ·· 150
6.4 DStream 实战 ·· 151
6.4.1 从端口读取数据 ·· 151
6.4.2 FileStream ·· 151
6.4.3 窗口函数 ·· 153
6.4.4 updateStateByKey ·· 154
6.5 Structured Streaming ··· 157
6.5.1 概述 ·· 157
6.5.2 快速示例 ·· 157

第 7 章 Spark ML 机器学习 ·· 161
7.1 机器学习 ··· 161
7.2 Spark ML ··· 163
7.3 典型机器学习流程介绍 ··· 163
7.3.1 提出问题 ·· 163
7.3.2 假设函数 ·· 164
7.3.3 损失函数 ·· 165
7.3.4 训练模型确定参数 ·· 166
7.4 经典算法模型实战 ··· 166
7.4.1 聚类算法实战 ·· 166
7.4.2 回归算法实战 ·· 170
7.4.3 协同过滤算法实战 ·· 172
7.4.4 分类算法实战 ·· 178

第 8 章 Spark GraphX 图计算 ·· 183
8.1 Spark GraphX ·· 183

8.2	Spark GraphX 的抽象	184
8.3	Spark GraphX 图的构建	185
8.4	Spark GraphX 图的计算模式	187
8.5	GraphX 3 个主要算法实战	189
8.6	GraphX 综合应用项目实战	192

第 9 章 Redis 数据库入门 ... 200

9.1	Redis 环境安装	200
	9.1.1 简介	200
	9.1.2 安装	201
	9.1.3 Java 客户端	202
9.2	Redis 常见数据类型	202
	9.2.1 key	202
	9.2.2 string 类型	204
	9.2.3 list	205
	9.2.4 set	206
	9.2.5 sorted set	208
	9.2.6 hash	209
9.3	Redis 排序	210
9.4	Redis 事务	213
9.5	Redis 发布订阅及示例	216
9.6	Redis 持久化	219

第 10 章 广告点击实时大数据分析项目实战 ... 221

10.1	项目环境准备	221
10.2	数据生成模块	226
10.3	从 Kafka 读取数据	230
	10.3.1 bean 类 AdsInfo	230
	10.3.2 工具类 MyKafkaUtil	230
	10.3.3 从 Kafka 消费数据	231
10.4	数据统计实现	233
	10.4.1 每天每地区热门广告点击率 Top3	233
	10.4.2 最近 1 小时内广告点击量实时统计	234

第 11 章 电影影评大数据分析项目实战 ... 237

11.1	项目介绍	237
11.2	项目实现	238
	11.2.1 公共代码开发	241
	11.2.2 平均评分最高的前 10 部电影	244
	11.2.3 电影类别及其平均评分	247
	11.2.4 评分次数最多的前 10 部电影	250

第 1 章

Spark 开发之 Scala 编程基础

本章详细讲解 Scala 的语法，包括基础语法、函数、控制语句、函数式编程、模式匹配、类和对象、异常处理、Trait（特征）、文件 I/O。掌握本章内容，可以为后续学习 Spark 数据分析奠定编程基础。注意：开发 Spark 代码和应用程序使用的 Scala 版本需要与 Spark 要求的 Scala 版本一致。

本章主要知识点：

- Scala 环境的安装
- Scala 基础语法
- Scala 函数与控制语句
- Scala 面向对象与 Trait 特性
- 异常处理与 I/O 流

1.1 开发环境搭建

可以在 IDEA 集成开发环境中使用 Scala、Java、Python 开发 Spark 应用。本节介绍使用 Scala 搭建 Spark 开发环境。

在启动 Spark（这里只是一个举例，具体可参看 2.2 节）时，会看到如下所示的信息，说明当前 Spark 所使用的 Scala 的版本为 2.13.8，JDK 的版本为 1.8.0。

```
Using Scala version 2.13.8 (Java HotSpot(TM) 64-Bit Server VM, Java 1.8.0_361)
Type in expressions to have them evaluated.
Type :help for more information.
```

开发 Scala 程序时，需要在本地安装 Scala 环境；如果使用 IDEA 开发环境，还需要在 IDEA 中安装 Scala 插件。安装 Scala 时，就像是安装 JDK 环境一样，也需要配置 Scala 的环境变量。

1. 安装 Scala

步骤 01 下载 Scala 安装包，后面 Spark 3.3.1 使用 Scala 2.13 版本，因此这里的下载地址如下：

```
https://www.scala-lang.org/download/2.13.8.html
https://downloads.lightbend.com/scala/2.13.8/scala-2.13.8.zip
```

解压并配置 SCALA_HOME 环境变量：

```
SCALA_HOME=D:\programfiles\scala-2.13.8
PATH=%SCALA_HOME%\bin
```

步骤 03 打开 CMD 命令行，查看 Scala 版本：

```
C:\>scala -version
Scala code runner version 2.13.8 -- Copyright 2002-2016, LAMP/EPFL and Lightbend,
```

步骤 04 运行 scala 命令，进入 Scala 命令行：

```
D:\a>scala
Welcome to Scala 2.13.8 (Java HotSpot(TM) 64-Bit Server VM, Java 1.8.0_361).
Type in expressions for evaluation. Or try :help.
scala> 1+1
res0: Int = 2
```

2. 在 IDEA 中安装 Scala 插件

检查自己安装的 IDEA 的版本，并安装对应的 Scala 插件，如图 1-1 所示。

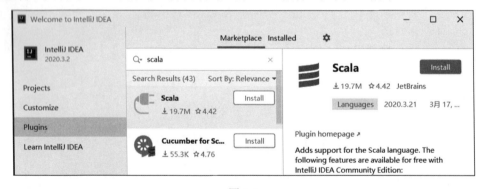

图 1-1

3. 开发 Spark 程序

步骤 01 在 IDEA 中创建项目模块 chapter1，添加 Scala 的支持，如图 1-2 和图 1-3 所示。

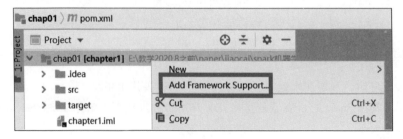

图 1-2

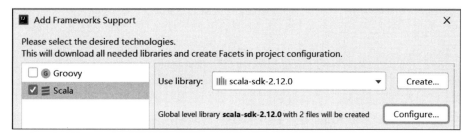

图 1-3

步骤 02 在 main 目录下创建 scala 目录，并设置为 resource root，如图 1-4 所示。

图 1-4

步骤 03 在 pom.xml 中添加依赖：

```
<dependency>
    <groupId>org.apache.spark</groupId>
    <artifactId>spark-core_2.12</artifactId>
    <version>3.3.1</version>
</dependency>
```

步骤 04 在 pom.xml 中添加编译 JDK 为 1.8 的插件（可选）：

```
<plugin>
    <groupId>org.apache.maven.plugins</groupId>
    <artifactId>maven-compiler-plugin</artifactId>
    <version>3.8.0</version>
    <configuration>
        <source>1.8</source>
        <target>1.8</target>
    </configuration>
</plugin>
```

步骤 05 在 pom.xml 中添加打包 Scala 到 JAR 文件中的插件：

```
<plugin>
    <groupId>net.alchim31.maven</groupId>
    <artifactId>scala-maven-plugin</artifactId>
    <version>4.4.1</version>
    <executions>
        <execution>
            <goals>
                <goal>compile</goal>
                <goal>testCompile</goal>
```

```
            </goals>
        </execution>
    </executions>
</plugin>
```

4. 测试 Scala 程序

步骤01 打开 IDEA，创建一个 Scala 程序 HelloScala，代码如下：

```
object HelloScala {
  def main(args: Array[String]): Unit = {
    println("Hello Scala")
  }
}
```

步骤02 在 IDEA 中直接运行 HelloScala 并输出结果：

```
Hello Scala
Process finished with exit code 0
```

步骤03 直接使用 maven 打包，如图 1-5 所示。

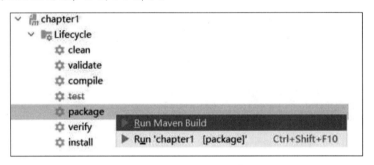

图 1-5

步骤04 将 JAR 文件放到任意目录下，因为需要 Scala 包的支持，所以使用 java -cp 运行 HelloScala 时必须在命令行上添加 Scala 的支持包再运行：

```
D:\a>java -cp spark-1.0.jar;%SCALA_HOME%\lib\* cn.isoft.HelloScala
Hello Scala..
```

1.2 基 础 语 法

如果之前学过 Java 语言并了解 Java 语言的基础知识，那么我们就能很快学会 Scala 的基础语法。Scala 与 Java 之间有些小的区别，比如 Scala 语句末尾的英文分号（;）是可选的。我们可以认为 Scala 程序是对象的集合，通过调用彼此的方法来实现消息传递。下面将详细介绍 Scala 编程语言的基础语法和编程常识。

1. 注释

注释有单行注释和多行注释。

```
// 单行注释开始于两个斜杠
/*
 * 多行注释，如之前所见，看起来像这样
 */
```

2. 打印

打印分两种：强制换行的打印和没有强制换行的打印。

```
//打印并强制换行
println("Hello world!")
println(10)
// 没有强制换行的打印
print("Hello world")
```

3. 变量

通过 var 或者 val 来声明变量。val 声明是不可变的，var 声明是可变的。不可变变量非常有用。

```
val x = 10        // x 现在是 10
x = 20            // 错误：对 val 声明的变量重新赋值
var y = 10
y = 20            // y 现在是 20
```

4. 数据类型

Scala 与 Java 有着相同的数据类型，表 1-1 列出了 Scala 支持的数据类型。

表 1-1 Scala 支持的数据类型

数 据 类 型	说　　明
Byte	8 位有符号补码整数。数值范围为-128~127
Short	16 位有符号补码整数。数值范围为-32768~ 32767
Int	32 位有符号补码整数。数值范围为-2147483648~2147483647
Long	64 位有符号补码整数。数值范围为-9223372036854775808~9223372036854775807
Float	32 位 IEEE754 标准的单精度浮点数
Double	64 位 IEEE754 标准的双精度浮点数
Char	16 位无符号 Unicode 字符，区间值为 U+0000~ U+FFFF
String	字符序列
Boolean	true 或 false
Unit	表示无值，和其他语言中 void 等同。用作不返回任何结果的方法的结果类型。Unit 只有一个实例值，写成()
Null	null 或空引用
Nothing	Nothing 类型在 Scala 的类层级的最底端；它是任何其他类型的子类型
Any	Any 是所有其他类的超类
AnyRef	AnyRef 类是 Scala 里所有引用类（reference class）的基类

Scala 数据类型设置示例如下：

```
val z: Int = 10
val a: Double = 1.0
```

t 到 Double 的自动转型，以下示例结果是 10.0，不是 10：

```
Double = 10.0
```

尔操作：

```
true              // false
!false            // true
true == false     // false
10 > 5            // true
```

5. 运算符

使用运算符进行数学运算：

```
1 + 1     // 2
2 - 1     // 1
5 * 3     // 15
6 / 2     // 3
6 / 4     // 1
6.0 / 4   // 1.5
```

6. 字符串

Scala 的字符串被英文双引号引起来，不存在单引号字符串。

String 有常见的 Java 字符串方法，例如：

```
"hello world".length
"hello world".substring(2, 6)
"hello world".replace("C", "3")
```

也有一些额外的 Scala 方法，例如：

```
"hello world".take(5)
"hello world".drop(5)
```

改写字符串时留意前缀"s"：

```
val n = 45
s"We have $n apples"  // => "We have 45 apples"
```

在要改写的字符串中使用表达式也是可以的：

```
val a = Array(11, 9, 6)
s"My second daughter is ${a(0) - a(2)} years old." // => "My second daughter is 5 years old"
s"We have double the amount of ${n / 2.0} in apples." // => "We have double the amount of 22.5 in apples."
s"Power of 2: ${math.pow(2, 2)}" // => "Power of 2: 4"
```

添加"f"前缀对要改写的字符串进行格式化：

```
f"Power of 5: ${math.pow(5, 2)}%1.0f" // "Power of 5: 25"
f"Square root of 122: ${math.sqrt(122)}%1.4f" //"Square root of 122: 11.0454"
```

未处理的字符串,忽略特殊字符:

```
raw"New line feed: \n. Carriage return: \r." // => "New line feed: n. Carriage return: r."
```

一些字符需要转义,比如字符串中的双引号:

```
"They stood outside the \"Rose and Crown\"" // => "They stood outside the "Rose and Crown""
```

三个双引号可以使字符串跨越多行,并包含引号:

```
val html = """<form id="daform">
<p>Press belo', Joe</p>
<input type="submit">
</form>"""
```

1.3 函　　数

函数是一组一起执行一个任务的语句。我们可以把代码划分到不同的函数中。如何划分代码到不同的函数中是我们自己来决定,但在逻辑上,划分通常是根据每个函数要执行一个特定的任务来进行的。

Scala 有函数和方法,二者在语义上的区别很小:Scala 方法是类的一部分,而函数是一个对象,可以赋值给一个变量。换句话来说,在类中定义的函数即是方法。

我们可以在任何地方定义函数,甚至可以在函数内定义函数(内嵌函数),更重要的一点是 Scala 函数名可以使用这些特殊字符:+、++、~、&、-、--、\、/、:等。

1. 函数声明

Scala 函数声明格式如下:

```
def functionName ([参数列表]) : [return type] { }
```

如果不写等于号和方法主体,那么方法会被隐式声明为"抽象"(abstract),于是包含它的类型也是一个抽象类型。

2. 函数定义

函数定义由一个 def 关键字开始,紧接着是可选的参数列表、一个英文冒号(:)、函数的返回类型、一个等于号(=),最后是函数的主体。

Scala 函数定义格式如下:

```
def functionName ([参数列表]) : [return type] = {
   function body
   return [expr]
}
```

其中 return type 可以是任意合法的 Scala 数据类型，参数列表中的参数可以使用逗号分隔。以下函数的功能是对两个传入的参数进行相加并求和：

```
object add{
  def addInt( a:Int, b:Int ) : Int = {
    var sum:Int = 0
    sum = a + b
    return sum
  }
}
```

如果函数没有返回值，那么可以返回 Unit，这个关键字类似于 Java 的 void，示例如下：

```
object Hello{
  def printMe( ) : Unit = {
    println("Hello, Scala!")
  }
}
```

3. 函数调用

Scala 提供了多种函数调用方式。

函数调用方法的标准格式如下：

```
functionName( 参数列表 )
```

如果函数使用了实例的对象来调用，那么我们可以使用类似 Java 的调用格式（使用"."号）：

```
[instance.]functionName( 参数列表 )
```

函数调用示例代码如下：

代码 1-1　TestFunc.scala

```
object TestFunc {
  def main(args: Array[String]) {
    println( "Returned Value : " + addInt(5,7) );
  }
  def addInt( a:Int, b:Int ) : Int = {
    var sum:Int = 0
    sum = a + b
    return sum
  }
}
```

执行以上代码，输出结果为：

```
Returned Value : 12
```

1.4 控制语句

1. 控制语句变量的使用

Scala 对点和括号的要求非常宽松（注意，它们的规则是不同的），这有助于写出读起来像英语的 DSL（领域特定语言）和 API（应用编程接口）。测试代码如代码 1-2 和代码 1-3 所示。

代码 1-2　Test Foreach.scala

```
1 to 5
val r = 1 to 5
r.foreach( println )
r foreach println
```

执行以上代码，输出结果为：

```
1,2,3,4,5,1
2
3
4
5
```

代码 1-3　Test Foreach2.scala

```
(5 to 1 by -1) foreach ( println )
```

执行以上代码，输出结果为：

```
5,4,3,2,1,
```

2. while 循环

while 循环是运行一系列语句，如果条件为 true，就重复运行，直到条件变为 false。测试代码如代码 1-4 所示。

代码 1-4　TestWhile.scala

```
var i = 0
while (i < 10) { println("i " + i); i+=1 }
```

执行以上代码，输出结果为：

```
i 0
i 1
i 2
i 3
i 4
i 5
i 6
i 7
i 8
i 9
```

3. do while 循环

do while 循环类似 while 语句，区别在于判断循环条件之前，do while 循环先执行一次循环的代码块。测试代码如代码 1-5 所示。

代码 1-5　TestDoWhile.scala

```
var x = 0;
do {
   println(x + " is still less than 10");
   x += 1
} while (x < 10)
```

执行以上代码，输出结果为：

```
0 is still less than 10
1 is still less than 10
2 is still less than 10
3 is still less than 10
4 is still less than 10
5 is still less than 10
6 is still less than 10
7 is still less than 10
8 is still less than 10
9 is still less than 10
```

4. for 循环

for 循环允许编写一个执行指定次数的循环控制结构。测试代码如代码 1-6 所示。

代码 1-6　TestFor.scala

```
def main(args: Array[String]) {
   var a = 0;
   // for 循环
   for( a <- 1 to 10){
      println( "Value of a: " + a );
   }
}
```

执行以上代码，输出结果为：

```
value of a: 1
value of a: 2
value of a: 3
value of a: 4
value of a: 5
value of a: 6
value of a: 7
value of a: 8
value of a: 9
value of a: 10
```

5. 条件语句

Scala 的 if...else 语句通过一条或多条语句的执行结果（True 或者 False）来决定执行的代码块。测试代码如代码 1-7 所示。

代码 1-7　Test If-else.scala

```
val x = 10
if (x == 1) println("yeah")
if (x == 10) println("yeah")
if (x == 11) println("yeah")
if (x == 11) println ("yeah") else println("nay")
println(if (x == 10) "yeah" else "nope")
val text = if (x == 10) "yeah" else "nope"
```

执行以上代码，输出结果为：

```
yeah
nay
yeah
```

6. break 语句

当在循环中使用 break 语句并执行到该语句时，就会中断循环并执行循环体之后的代码块。Scala 语言中默认是没有 break 语句的，但是在 Scala 2.8 版本后可以使用另外一种方式来实现 break 语句。

Scala 中 break 的语法格式如下：

```
// 导入以下包
import scala.util.control._
// 创建 Breaks 对象
val loop = new Breaks;
// 在 breakable 中循环
loop.breakable{
    // 循环
    for(...){
        ...
        // 循环中断
        loop.break;
    }
}
```

测试代码如代码 1-8 所示。

代码 1-8　TestBreak.scala

```
import scala.util.control._
object TestBreak {
  def main(args: Array[String]) {
    var a = 0;
    val numList = List(1,2,3,4,5,6,7,8,9,10);

    val loop = new Breaks;
    loop.breakable {
```

```
        for( a <- numList){
          println( "Value of a: " + a );
          if( a == 4 ){
            loop.break;
          }
        }
      }
      println( "After the loop" );
    }
}
```

执行以上代码，输出结果为：

```
Value of a: 1
Value of a: 2
Value of a: 3
Value of a: 4
After the loop
```

1.5 函数式编程

1. Array（数组）

Scala 数组声明的语法格式如下：

```
var z:Array[String] = new Array[String](3)
或
var z = new Array[String](3)
```

数组的元素类型和数组的大小都是确定的，所以当处理数组元素时，我们通常使用基本的 for 循环来遍历数组元素。

以下示例演示了数组的创建、初始化等处理过程。

代码 1-9　TestArray1.scala

```
object TestArray1 {
  def main(args: Array[String]) {
    var myList = Array(1.9, 2.9, 3.4, 3.5)
    // 输出所有数组元素
    for ( x <- myList ) {
      println( x )
    }
    // 计算数组所有元素的总和
    var total = 0.0;
    for ( i <- 0 to (myList.length - 1)) {
      total += myList(i);
    }
    println("总和为 " + total);
    // 查找数组中的最大元素
```

```
        var max = myList(0);
        for ( i <- 1 to (myList.length - 1) ) {
            if (myList(i) > max) max = myList(i);
        }
        println("最大值为 " + max);
    }
}
```

执行以上代码,输出结果为:

```
1.9
2.9
3.4
3.5
总和为 11.7
最大值为 3.5
```

2. List(列表)

List 的特征是其元素以线性方式存储,列表中可以存放重复对象。

以下列出了多种类型的列表:

```
// 字符串列表
val site: List[String] = List("mrchi 的博客", "Google", "Baidu")
// 整型列表
val nums: List[Int] = List(1, 2, 3, 4)
// 空列表
val empty: List[Nothing] = List()
// 二维列表
val dim: List[List[Int]] =
    List(
        List(1, 0, 0),
        List(0, 1, 0),
        List(0, 0, 1)
    )
```

对于 Scala 列表的任何操作都可以使用 head、tail、isEmpty 这 3 个基本操作来表达,示例如下:

代码 1-10   TestList.scala

```
object TestList {
    def main(args: Array[String]) {
        val site = "mrchi 的博客" :: ("Google" :: ("Baidu" :: Nil))
        val nums = Nil
        println( "第一网站是 : " + site.head )
        println( "最后一个网站是 : " + site.tail )
        println( "查看列表 site 是否为空 : " + site.isEmpty )
        println( "查看 nums 是否为空 : " + nums.isEmpty )
    }
}
```

执行以上代码,输出结果为:

```
第一网站是：mrchi 的博客
最后一个网站是：List(Google, Baidu)
查看列表 site 是否为空：false
查看 nums 是否为空：true
```

3. Set（集合）

Set 是最简单的一种集合。集合中的对象不按特定的方式排序，并且没有重复对象。

对于 Scala 集合的任何操作都可以使用 head、tail、isEmpty 这 3 个基本操作来表达，示例如下：

代码 1-11　TestSet.scala

```scala
object TestSet {
  def main(args: Array[String]) {
    val site = Set("mrchi的博客", "Google", "Baidu")
    val nums: Set[Int] = Set()
    println( "第一网站是: " + site.head )
    println( "最后一个网站是: " + site.tail )
    println( "查看列表site是否为空: " + site.isEmpty )
    println( "查看 nums 是否为空: " + nums.isEmpty )
  }
}
```

执行以上代码，输出结果为：

```
第一网站是：mrchi 的博客
最后一个网站是：Set(Google, Baidu)
查看列表 site 是否为空：false
查看 nums 是否为空：true
```

4. Map（映射）

Map 是一种映射键对象和值对象的集合，它的每一个元素都包含一对键对象和值对象。

以下示例演示 key、values、isEmpty 这 3 个方法的基本应用。

代码 1-12　TestMap.scala

```scala
object TestMap {
  def main(args: Array[String]) {
    val colors = Map("red" -> "#FF0000",
                     "azure" -> "#F0FFFF",
                     "peru" -> "#CD853F")
    val nums: Map[Int, Int] = Map()
    println( "colors 中的键为: " + colors.keys )
    println( "colors 中的值为: " + colors.values )
    println( "检测 colors 是否为空: " + colors.isEmpty )
    println( "检测 nums 是否为空: " + nums.isEmpty )
  }
}
```

执行以上代码，输出结果为：

```
colors 中的键为：Set(red, azure, peru)
```

colors 中的值为：MapLike(#FF0000, #F0FFFF, #CD853F)
检测 colors 是否为空：false
检测 nums 是否为空：true

5. 元组

元组是不同类型的值的集合。与列表一样，元组也是不可变的，但与列表不同的是元组可以包含不同类型的元素。

元组的值是通过将单个的值包含在圆括号中构成的。例如：

```
val t = (1, 3.14, "Fred")
```

表示在元组中定义了 3 个元素，对应的类型分别为[Int, Double, java.lang.String]。

此外也可以使用以下方式来定义元组：

```
val t = new Tuple3(1, 3.14, "Fred")
```

可以使用 t._1 访问第一个元素，t._2 访问第二个元素，以此类推。元组的示例代码如下：

代码 1-13　TestTuple.scala

```
object TestTuple {
  def main(args: Array[String]) {
    val t = (4,3,2,1)
    val sum = t._1 + t._2 + t._3 + t._4
    println( "元素之和为: " + sum )
  }
}
```

执行以上代码，输出结果为：

元素之和为：10

6. Option

Option[T] 表示有可能包含值的容器，当然也可能不包含值。Scala Iterator（迭代器）不是一个容器，更确切地说它是逐一访问容器内元素的方法。Scala Option（选项）类型用来表示一个值是可选的（有值或无值）。

Option[T] 是一个类型为 T 的可选值的容器：如果值存在，那么 Option[T]就是一个 Some[T]；如果不存在，那么 Option[T] 就是对象 None 。

接下来我们来看一段代码：

```
// 虽然 Scala 可以不定义变量的类型，不过为了清楚些，还是把它显示地定义上
val myMap: Map[String, String] = Map("key1" -> "value")
val value1: Option[String] = myMap.get("key1")
val value2: Option[String] = myMap.get("key2")
println(value1) // Some("value1")
println(value2) // None
```

代码解释：

（1）在上面的代码中，myMap 是一个键的类型是 String、值的类型是 String 的 hash map，但

不一样的是它的 get()返回的是一个叫作 Option[String]的类别。

（2）Scala 使用 Option[String]来告诉我们："我会想办法回传一个 String，但也可能没有 String 给你"。

（3）myMap 里并没有 key2 数据，因此 get()方法返回 None。

Option 有两个子类别，一个是 Some，一个是 None：当它回传 Some 的时候，代表这个函数成功地给了我们一个 String，而我们可以通过 get()函数拿到那个 String；如果它返回的是 None，则代表没有字符串可以给我们。示例代码如下：

代码 1-14　TestOption.scala

```
object Test {
  def main(args: Array[String]) {
    val sites = Map("余辉" -> "mrchi的博客", "google" -> "www.google.com")
    println("sites.get( \"余辉\" ) : " + sites.get( "余辉" )) // Some(www.runoob.com)
    println("sites.get( \"baidu\" ) : " + sites.get( "baidu" ))  // None
  }
}
```

执行以上代码，输出结果为：

```
sites.get( "runoob" ) : Some(mrchi的博客)
sites.get( "baidu" ) : None
```

也可以通过模式匹配来输出匹配值，示例代码如下：

代码 1-15　TestOption2.scala

```
object Test {
  def main(args: Array[String]) {
    val sites = Map("余辉" -> "mrchi的博客", "google" -> "www.google.com")
    println("show(sites.get( \"余辉\")) : " +
      show(sites.get( "余辉")) )
    println("show(sites.get( \"baidu\")) : " +
      show(sites.get( "baidu")) )
  }
  def show(x: Option[String]) = x match {
    case Some(s) => s
    case None => "?"
  }
}
```

执行以上代码，输出结果为：

```
show(sites.get( "余辉")) : mrchi的博客
show(sites.get( "baidu")) : ?
```

1.6 模 式 匹 配

1. 模式匹配

Scala 提供了强大的模式匹配机制，应用得也非常广泛。

一个模式匹配包含了一系列备选项，每个备选项都开始于关键字 case，包含了一个模式及一到多个表达式。箭头符号（=>）隔开了模式和表达式。

以下是一个简单的整型值模式匹配示例代码。

代码 1-16　TestMach.scala

```
object Test {
   def main(args: Array[String]) {
      println(matchTest(3))
   }
   def matchTest(x: Int): String = x match {
      case 1 => "one"
      case 2 => "two"
      case _ => "many"
   }
}
```

执行以上代码，输出结果为：

```
many
```

示例代码 match 对应 Java 里的 switch，但是写在选择器表达式之后，即选择器 match{备选项}。

match 表达式通过按照代码编写的先后次序尝试匹配每个模式来完成计算，只要发现有一个匹配的 case，剩下的 case 就不会继续匹配。

接下来我们来看一个不同数据类型的模式匹配示例代码。

代码 1-17　TestPattern.scala

```
object TestPattern {
   def main(args: Array[String]) {
      println(matchTest("two"))
      println(matchTest("test"))
      println(matchTest(1))
      println(matchTest(6))
   }
   def matchTest(x: Any): Any = x match {
      case 1 => "one"
      case "two" => 2
      case y: Int => "scala.Int"
      case _ => "many"
   }
}
```

执行以上代码，输出结果为：

```
2
many
one
scala.Int
```

代码解析：

第 1 个 case 对应整型数值 1；第 2 个 case 对应字符串值"two"；第 3 个 case 对应类型模式，用于判断传入的值是否为整型，相比使用 isInstanceOf 来判断类型，使用模式匹配更好；第 4 个 case 表示默认的全匹配备选项，即没有找到其他匹配时的匹配项，类似 switch 中的 default。

2. 样例类

使用了 case 关键字的类定义就是样例类（case classes），样例类是一种特殊的类，经过优化后用于模式匹配。以下是样例类的简单示例代码。

代码 1-18　TestPattern1.scala

```
object TestPattern1 {
  def main(args: Array[String]) {
    val alice = new Person("Alice", 25)
    val bob = new Person("Bob", 32)
    val charlie = new Person("Charlie", 32)
    for (person <- List(alice, bob, charlie)) {
    person match {
        case Person("Alice", 25) => println("Hi Alice!")
        case Person("Bob", 32) => println("Hi Bob!")
        case Person(name, age) =>
          println("Age: " + age + " year, name: " + name + "?")
      }
    }
  }
  // 样例类
  case class Person(name: String, age: Int)
}
```

执行以上代码，输出结果为：

```
Hi Alice!
Hi Bob!
Age: 32 year, name: Charlie?
```

1.7　类 和 对 象

1. 类的定义

类是对象的抽象，而对象是类的具体实例。类是抽象的，不占用内存，而对象是具体的，占用存储空间。类是用于创建对象的蓝图，是一个定义许多具有共性特征和行为的对象的软件模板。

Scala 中的类不声明为 public，一个 Scala 源文件中可以有多个类。我们可以使用 new 关键字来创建类的对象，示例如下：

```
class Point(xc: Int, yc: Int) {
  var x: Int = xc
  var y: Int = yc
  def move(dx: Int, dy: Int) {
    x = x + dx
    y = y + dy
    println ("x 的坐标点: " + x);
    println ("y 的坐标点: " + y);
  }
}
```

代码解析：

示例中类定义了两个变量 x 和 y；还定义了一个方法 move，方法没有返回值。

Scala 的类定义可以有参数，称之为类参数，如上述示例中的 xc、yc，类参数在整个类中都可以访问。使用 new 来实例化类并访问类中的方法和变量的示例，如代码 1-19 所示。

代码 1-19　TestPoint.scala

```
import java.io._
class Point(xc: Int, yc: Int) {
  var x: Int = xc
  var y: Int = yc
  def move(dx: Int, dy: Int) {
    x = x + dx
    y = y + dy
    println ("x 的坐标点: " + x);
    println ("y 的坐标点: " + y);
  }
}
object TestPoint {
  def main(args: Array[String]) {
    val pt = new Point(10, 20);

    // 移到一个新的位置
    pt.move(10, 10);
  }
}
```

执行以上代码，输出结果为：

x 的坐标点: 20
y 的坐标点: 30

2. 继承

Scala 使用 extends 关键字来继承一个类。Scala 继承一个基类跟 Java 很相似，但需要注意以下几点：

（1）重写一个非抽象方法时必须使用 override 修饰符。

（2）只有主构造函数才可以往基类的构造函数里写参数。

（3）在子类中重写超类的抽象方法时，不需要使用 override 关键字。

接下来让我们来看个示例。

代码 1-20　TestInherit.scala

```
class Point(xc: Int, yc: Int) {
  var x: Int = xc
  var y: Int = yc
  def move(dx: Int, dy: Int) {
    x = x + dx
    y = y + dy
    println ("x 的坐标点: " + x);
    println ("y 的坐标点: " + y);
  }
}
class Location(override val xc: Int, override val yc: Int,
  val zc :Int) extends Point(xc, yc){
  var z: Int = zc
  def move(dx: Int, dy: Int, dz: Int) {
    x = x + dx
    y = y + dy
    z = z + dz
    println ("x 的坐标点 : " + x);
    println ("y 的坐标点 : " + y);
    println ("z 的坐标点 : " + z);
  }
}
```

代码解析：

示例中 Location 类继承了 Point 类，Point 称为父类（基类），Location 称为子类。override val xc 为重写了父类的字段。

继承会继承父类的所有属性和方法，Scala 只允许继承一个父类。示例代码如下：

代码 1-21　TestInherit2.scala

```
import java.io._
class Point(val xc: Int, val yc: Int) {
  var x: Int = xc
  var y: Int = yc
  def move(dx: Int, dy: Int) {
    x = x + dx
    y = y + dy
    println ("x 的坐标点 : " + x);
    println ("y 的坐标点 : " + y);
  }
}
class Location(override val xc: Int, override val yc: Int,
  val zc :Int) extends Point(xc, yc){
```

```
    var z: Int = zc

    def move(dx: Int, dy: Int, dz: Int) {
      x = x + dx
      y = y + dy
      z = z + dz
      println ("x 的坐标点 : " + x);
      println ("y 的坐标点 : " + y);
      println ("z 的坐标点 : " + z);
    }
}
object Test {
  def main(args: Array[String]) {
    val loc = new Location(10, 20, 15);

    // 移到一个新的位置
    loc.move(10, 10, 5);
  }
}
```

执行以上代码，输出结果为：

```
x 的坐标点 : 20
y 的坐标点 : 30
z 的坐标点 : 20
```

Scala 重写一个非抽象方法时，必须用 override 修饰符。示例代码如下：

代码 1-22　TestInherit3.scala

```
class Person {
  var name = ""
  override def toString = getClass.getName + "[name=" + name + "]"
}
class Employee extends Person {
  var salary = 0.0
  override def toString = super.toString + "[salary=" + salary + "]"
}
object TestInherit1 extends App {
  val fred = new Employee
  fred.name = "Fred"
  fred.salary = 50000
  println(fred)
}
```

执行以上代码，输出结果为：

```
Employee[name=Fred][salary=50000.0]
```

1.8 异常处理

Scala 的异常处理与其他语言（比如 Java）类似。Scala 可以通过抛出异常的方式来终止相关代码的运行，而不必通过返回值。

1. 抛出异常

Scala 抛出异常的方法和 Java 一样，使用 throw 方法。例如，抛出一个非法参数异常：

```
throw new IllegalArgumentException
```

2. 捕获异常

Scala 异常捕捉的机制与其他语言的处理方法一样，如果有异常发生，那么 catch 子句按次序捕捉。因此，在 catch 子句中，越具体的异常越靠前，越普遍的异常越靠后。如果抛出的异常不在 catch 子句中，该异常则无法处理，会被升级到调用者处。

捕捉异常的 catch 子句的语法与其他语言中的不太一样。在 Scala 里，借用了模式匹配的思想来做异常的匹配，因此，在 catch 的代码里是一系列 case 字句，如代码 1-23 所示。

代码 1-23　TestException.scala

```
import java.io.FileReader
import java.io.FileNotFoundException
import java.io.IOException
object Test {
  def main(args: Array[String]) {
    try {
      val f = new FileReader("input.txt")
    } catch {
      case ex: FileNotFoundException =>{
        println("Missing file exception")
      }
      case ex: IOException => {
        println("IO Exception")
      }
    }
  }
}
```

执行以上代码，输出结果为：

```
Missing file exception
```

catch 语句里的内容跟 match 里的 case 是完全一样的。由于异常捕捉是按次序的，如果把最普遍的异常 Throwable 写在最前面，则在它后面的 case 都捕捉不到，因此需要将它写在最后面。

3. finally 语句

finally 语句用于执行不管是正常处理还是有异常发生时都需要执行的步骤，如代码 1-24 所示。

代码1-24　TestFinally.scala

```scala
import java.io.FileReader
import java.io.FileNotFoundException
import java.io.IOException
object TestFinally {
  def main(args: Array[String]) {
    try {
      val f = new FileReader("input.txt")
    } catch {
      case ex: FileNotFoundException => {
        println("Missing file exception")
      }
      case ex: IOException => {
        println("IO Exception")
      }
    } finally {
      println("Exiting finally...")
    }
  }
}
```

执行以上代码，输出结果为：

```
Missing file exception
Exiting finally...
```

1.9　Trait（特征）

Scala 的 Trait（特征）相当于 Java 的接口，但它比接口的功能还要强大，它还可以定义属性和方法的实现。

一般情况下 Scala 的类只能够继承单一父类，但是如果是 Trait 的话就可以继承多个，从结果来看就是实现了多重继承。

Trait 定义的方式与类类似，但它使用的关键字是 trait，示例如下：

```scala
trait Equal {
  def isEqual(x: Any): Boolean
  def isNotEqual(x: Any): Boolean = !isEqual(x)
}
```

代码解析：

以上 Trait 由两个方法组成：isEqual 和 isNotEqual。isEqual 方法没有定义方法的实现，isNotEqual 定义了方法的实现。

子类继承特征可以实现未被实现的方法，所以 Scala Trait 其实更像 Java 的抽象类。

特征的完整示例如代码 1-25 所示。

代码 1-25　TestTrait.scala

```scala
trait Equal {
  def isEqual(x: Any): Boolean
  def isNotEqual(x: Any): Boolean = !isEqual(x)
}
class Point(xc: Int, yc: Int) extends Equal {
  var x: Int = xc
  var y: Int = yc
  def isEqual(obj: Any) =
    obj.isInstanceOf[Point] &&
    obj.asInstanceOf[Point].x == x
}
object Test {
  def main(args: Array[String]) {
    val p1 = new Point(2, 3)
    val p2 = new Point(2, 4)
    val p3 = new Point(3, 3)
    println(p1.isNotEqual(p2))
    println(p1.isNotEqual(p3))
    println(p1.isNotEqual(2))
  }
}
```

执行以上代码，输出结果为：

```
false
true
true
```

1.10　文件 I/O

1. I/O 介绍

Scala 进行文件写操作直接使用的是 Java 中的 I/O 类（java.io.File），如代码 1-26 所示。

代码 1-26　TestFileWriter.scala

```scala
import java.io._
object TestFileWriter {
  def main(args: Array[String]) {
    val writer = new PrintWriter(new File("test.txt" ))
    writer.write("博客地址为 http://blog.csdn.net/mrchi")
    writer.close()
  }
}
```

执行以上代码，会在当前目录下生成一个 test.txt 文件，文件内容为"mrchi 的博客 http://blog.csdn.net/silentwolfyh"。

2. 从屏幕上读取用户输入

有时候我们需要接收用户在屏幕上输入的指令来处理程序,如代码 1-27 所示。

代码 1-27　TestScreenRead.scala

```scala
object Test {
  def main(args: Array[String]) {
    print("请输入博客地址: " )
    val line = Console.readLine
    println("谢谢,你输入的是: " + line)
  }
}
```

执行以上代码,屏幕上会显示如下信息:

请输入博客地址: `http://blog.csdn.net/mrchi`
谢谢,你输入的是: `http://blog.csdn.net/mrchi`

3. 从文件上读取内容

从文件读取内容非常简单,我们可以使用 Scala 的 Source 类及伴生对象来读取文件。代码 1-28 演示的是从 "test.txt"(代码 1-26 中创建)文件中读取内容。

代码 1-28　TestFileRead.scala

```scala
import scala.io.Source
object TestFileRead {
  def main(args: Array[String]) {
    println("文件内容为:" )
    Source.fromFile("test.txt" ).foreach{
      print
    }
  }
}
```

执行以上代码,输出结果为:

文件内容为:博客地址为 `http://blog.csdn.net/mrchi`

第 2 章

Spark 框架全生态体验

本章主要介绍 Spark 的基本概念、Spark 不同运行模式的安装和 Spark 关键技术三大部分内容。读者通过本章可以对 Spark 框架及其关键技术有一个全面的了解。

本章主要知识点：

- Spark 概述
- Linux 环境搭建
- Hadoop 安装与配置
- Spark 安装与配置
- spark-submit 的使用
- DataFrame 初体验
- Spark SQL 初体验
- Spark Streaming 初体验
- 共享变量

2.1　Spark 概述

本节主要介绍 Spark 框架及其基本概念。

2.1.1　关于 Spark

Apache Spark 是专为大规模数据处理而设计的快速通用的计算引擎。

Spark 是加州大学伯克利分校的 AMP 实验室（Algorithms，Machines and People Lab）开源的类 Hadoop MapReduce 的通用并行框架，拥有 Hadoop MapReduce 所具有的优点，但不同于 MapReduce 的是 Job（工作）中间输出的结果可以保存在内存中，从而不再需要读写 HDFS，因此 Spark 能更好地适用于数据挖掘与机器学习等需要迭代的 MapReduce 的算法。

Spark 是一种与 Hadoop 相似的开源集群计算环境，但是两者之间存在一些不同之处，这些不同之处使 Spark 在某些工作负载方面表现得更加优越，换句话说，Spark 启用了内存分布数据集，除了能够提供交互式查询外，它还可以优化迭代工作负载。

Spark 是在 Scala 语言中实现的，它将 Scala 用作应用程序框架。与 Hadoop 不同，Spark 和 Scala 能够紧密集成，其中的 Scala 可以像操作本地集合对象一样轻松地操作分布式数据集。

尽管创建 Spark 是为了支持分布式数据集上的迭代作业，但是实际上它是对 Hadoop 的补充，可以在 Hadoop 文件系统中并行运行。通过名为 Mesos 的第三方集群框架可以支持此行为。Spark 可以用来构建大型的、低延迟的数据分析应用程序。

2.1.2 Spark 的基本概念

1. Spark 特性

Spark 具有以下特性：

- 高可伸缩性。
- 高容错。
- 内存计算。

2. Spark 的生态体系

Spark 属于 BDAS（伯利克分析栈）生态体系。

- MapReduce 属于 Hadoop 生态体系之一，Spark 则属于 BDAS 生态体系之一。
- Hadoop 包含了 MapReduce、HDFS、HBase、Hive、ZooKeeper、Pig、Sqoop 等。
- BDAS 包含了 Spark GraphX、Spark SQL（相当于 Hive）、Spark MLlib、Spark Streaming（消息实时处理框架，类似 Storm）、BlinkDB 等。

BDAS 生态体系图如图 2-1 所示。

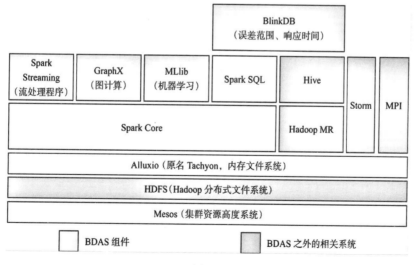

图 2-1

3. Spark 与 MapReduce

相对于 MapReduce，Spark 具有如下优势：

- MapReduce 通常将中间结果存放到 HDFS 上，Spark 则是基于内存并行大数据框架，中间结果存放到内存。对于迭代数据而言 Spark 效率更高。
- MapReduce 总是消耗大量时间排序，而有些场景不需要排序，Spark 可以避免不必要的排序所带来的开销。
- Spark 是一张有向无环图（从一个点出发最终无法回到该点的一个拓扑），并对有向无环图对应的流程进行优化。

Spark 为什么比 MapReduce 快？简单说，Spark①基于内存计算、减少了低效的磁盘交互；②使用基于 DAG 的高效调度算法；③具有容错机制 Linage（血统）。

4. Spark 支持的 API

Spark 支持的 API 包括 Scala、SQL、Python、Java、R 等。

5. Spark 运行模式

Spark 有 5 种运行模式，其中 Local 是单机模式，其他 4 种都是集群模式：

- Local：Spark 运行在本地模式上，用于测试、开发。本地模式就是以一个独立的进程，通过其内部的多个线程来模拟整个 Spark 运行时环境。
- Standlone：Spark 运行在独立集群模式上。Spark 中的各个角色以独立进程的形式存在，并组成 Spark 集群环境。
- Hadoop YARN：Spark 运行在 YARN 上。Spark 中的各个角色运行在 YARN 的容器内部，并组成 Spark 集群环境。
- Apache Mesos：Spark 中的各个角色运行在 Apache Mesos 上，并组成 Spark 集群环境。
- Kubernetes：Spark 中的各个角色运行在 Kubernetes 的容器内部，并组成 Spark 集群环境。

2.1.3 Spark 集群模式

本小节简单讲解一下 Spark 如何在集群上运行，以便更容易理解所涉及的相关组件。

1. Spark 集群的组件

Spark 应用程序（Application）在集群上作为独立的进程集运行，由驱动程序（Driver Program，又称主程序）中的 SparkContext 对象进行协调。具体来说，Spark 应用程序要在集群上运行，SparkContext 可以连接到几种类型的集群管理器（Cluster Manager，包括 Spark 自己的独立集群管理器、Mesos、YARN 或 Kubernetes），这些集群管理器在应用程序之间分配资源。连接后，Spark 会获取集群中节点上的执行器（Executor），这些节点是为应用程序运行计算和存储数据的进程。接下来，它将应用程序代码（由传递给 SparkContext 的 JAR 或 Python 文件定义）发送给执行器。最后，SparkContext 将任务发送给执行器以运行。Spark 集群组件如图 2-2 所示。

Driver Program（驱动程序）启动多个 Worker Node，Worker 从文件系统加载数据并产生 RDD

（即数据存放到 RDD 中，RDD 是一个数据结构），再按照不同分区缓存到内存中。

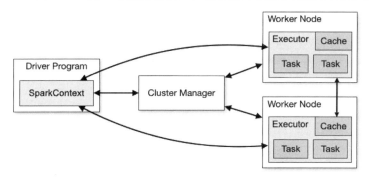

图 2-2

这个架构中有几个要点需要注意：

（1）每个应用程序都有自己的执行器进程，这些进程在整个应用程序期间保持运行，并在多个线程中运行任务。这样做的好处是在调度端（每个驱动程序调度自己的任务）和执行器端（来自不同应用程序的任务在不同 JVM 中运行）将应用程序彼此隔离。然而，这也意味着，如果不将数据写入外部存储系统，就无法在不同的 Spark 应用程序（SparkContext 实例）之间共享数据。

（2）Spark 对底层集群管理器是不可知的。只要它可以获取执行器进程，并且这些进程相互通信，那么即使在其他应用程序（例如 Mesos/YARN/Kubernetes）的集群管理器上运行它也相对容易。

（3）驱动程序必须在其整个生命周期中监听并接收来自执行程序的传入连接（请参阅网络配置部分中的 spark.driver.port），因此，驱动程序必须是可从工作节点进行网络寻址的。

（4）因为驱动程序在集群上调度任务，所以它应该在工作节点附近运行，最好在同一局域网上运行。如果想远程向集群发送请求，最好打开一个 RPC 到驱动程序，让它从附近节点提交操作（指 transformation 和 action），而不是在远离工作节点的地方运行驱动程序。

2. 集群管理器类型

Spark 当前支持以下几种集群管理器：

- Standalone：Spark 附带的一个简单的集群管理器，可以轻松地设置集群。
- Apache Mesos：一个通用的集群管理器，也可以运行 HadoopMapReduce 和服务应用程序（已弃用）。
- Hadoop YARN：Hadoop 2 和 Hadoop 3 中的资源管理器。
- Kubernetes：一个用于自动化容器化应用程序的部署、扩展和管理的开源系统。

3. 作业安排

Spark 可以控制应用程序之间（在集群管理器级别）和应用程序内部（如果在同一 SparkContext 上进行多个计算）的资源分配。

4. Spark 集群常用术语

Spark 集群常用术语如表 2-1 所示。

表2-1　Spark集群常用术语

术　　语	含　　义
Application	基于 Spark 的用户程序，包含一个 Driver Program 和集群中的多个 Executor
Application jar	一个包含用户 Spark 应用程序的 JAR 包。在某些情况下，用户希望创建一个 "uber jar"，其中包含他们的应用程序及其依赖项。用户的 JAR 包永远不包括 Hadoop 或 Spark 库，但是这些库需要在 JAR 包运行时添加上
Driver program	运行 Application 的 main() 函数并创建 SparkContext 的进程
Cluster manager	在集群上获取资源的外部服务，例如：Standalone、Mesos 或 YARN、Kubernetes 等集群管理系统
Deploy mode	区分驱动程序进程运行的位置。在"集群"模式下，框架在集群内部启动驱动程序；在"客户端"模式下，提交者在集群之外启动驱动程序
Worker node	集群中运行 Application 的任何节点
Executor	是为某 Application 运行在 Worker Node 上的一个进程，该进程负责运行 Task，并且负责将数据存储在内存或者磁盘上，每个 Application 都有各自独立的 Executor
Task	被送到一个 Executor 上的工作单元
Job	由多个任务组成的并行计算，这些任务是响应 Spark 操作（例如保存、收集）而派生的；我们会在驱动程序日志中看到这个术语
Stage	每个作业被划分为更小的任务集，称为相互依赖的阶段（类似于 MapReduce 中的 map 和 reduce 阶段）；我们会在驱动程序日志中看到这个术语

5. RDD

RDD 英文名为 Resilient Distributed Dataset，中文名翻译为弹性分布式数据集。

什么是 RDD？RDD 是一个只读、分区记录的集合，我们可以把它理解为一个存储数据的数据结构。也就是说，RDD 是 Spark 对数据的核心抽象，其实就是分布式的元素集合。在 Spark 中，对数据的所有操作不外乎创建 RDD、转化已有 RDD 以及调用 RDD 操作进行求值。而在这一切操作背后，Spark 会自动将 RDD 中的数据分发到集群上，并将操作并行化执行。在 Spark 中的一切操作都是基于 RDD 的。

RDD 可以通过以下 3 种方式创建：

- 集合转换。
- 从文件系统（本地文件、HDFS、HBase）输入。
- 从父 RDD 转换（为什么需要父 RDD 呢？为了容错）。

RDD 的计算类型有以下 2 种：

- Transformation: 延迟执行。一个 RDD 通过该操作产生新的 RDD 时不会立即执行，只有等到 Action 操作才会真正执行。
- Action: 提交 Spark 作业。当执行 Action 时，Transformation 类型的操作才会真正执行计算操作，然后产生最终结果并输出。

Hadoop 提供处理的数据接口有 Map 和 Reduce，而 Spark 提供的不仅仅有 Map 和 Reduce，还有更多对数据处理的接口。Spark 算子包括转换算子和行动算子，这部分内容将在 3.4 节进行集中讨论。

6. 容错

每个 RDD 都会记录自己所依赖的父 RDD，一旦出现某个 RDD 的某些 Partition（分区）丢失，可以通过并行计算迅速恢复，这就是容错。

RDD 的依赖又分为 Narrow Dependent（窄依赖）和 Wide Dependent（宽依赖）。

（1）窄依赖：每个 Partition 最多只能给一个 RDD 使用，由于没有多重依赖，所以在一个节点上可以一次性将 Partition 处理完，且一旦数据发生丢失或者损坏，可以迅速从上一个 RDD 恢复。

（2）宽依赖：每个 Partition 可以给多个 RDD 使用，由于多重依赖，只有等到所有到达节点的数据处理完毕才能进行下一步处理，一旦发生数据丢失或者损坏，需要从所有父 RDD 重新计算，相对窄依赖而言付出的代价更高。所以在此发生之前，必须将上一次所有节点的数据进行物化（存储到磁盘上）处理，这样达到恢复，因此也应该尽量避免宽依赖的使用。

RDD 宽、窄依赖如图 2-3 所示。

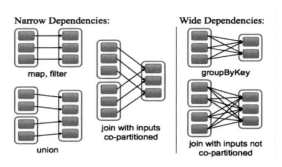

图 2-3

7. 缓存策略

Spark 通过 useDisk、useMemory、useOffHeap、deserialized、replication 5 个参数组成 11 种缓存策略。这 11 种缓存策略如下：

（1）DISK_ONLY。

参数：_useDisk, _useMemory, _useOffHeap, _deserialized, _replication（默认值为 1）。

（2）DISK_ONLY_2：副本 2 份。

参数：_useDisk, _useMemory, _useOffHeap, _deserialized, _replication（默认值为 1）。

（3）MEMORY_ONLY：默认的。

参数：_useDisk, _useMemory, _useOffHeap, _deserialized, _replication（默认值为 1）。

（4）MEMORY_ONLY_2。

参数：_useDisk, _useMemory, _useOffHeap, _deserialized, _replication（默认值为 1）。

（5）MEMORY_ONLY_SER：SER 做序列化，会消耗 CPU。

参数：_useDisk, _useMemory, _useOffHeap, _deserialized, _replication（默认值为 1）。

（6）MEMORY_ONLY_SER_2。

参数：_useDisk, _useMemory, _useOffHeap, _deserialized, _replication（默认值为1）。

（7）MEMORY_AND_DISK：内存中若放不下，则多出的部分放在机器的本地磁盘上。

参数：_useDisk, _useMemory, _useOffHeap, _deserialized, _replication（默认值为1）。

（8）MEMORY_AND_DISK_2。

参数：_useDisk, _useMemory, _useOffHeap, _deserialized, _replication（默认值为1）。

（9）MEMORY_AND_DISK_SER。

参数：_useDisk, _useMemory, _useOffHeap, _deserialized, _replication（默认值为1）。

（10）MEMORY_AND_DISK_SER_2。

参数：_useDisk, _useMemory, _useOffHeap, _deserialized, _replication（默认值为1）。

（11）OFF_HEAP：不使用堆，比如可以使用Tachyon。

参数：_useDisk, _useMemory, _useOffHeap, _deserialized, _replication（默认值为1）、NONE。

参数说明如下：

- useDisk：使用磁盘缓存（boolean）。
- useMemory：使用内存缓存（boolean）。
- useOffHeap：是否使用Java的堆外内存。
- deserialized：反序列化（序列化是方便数据在网络中以对象的形式进行传输，boolean：true表示反序列化，false表示序列化）。
- replication：副本数量（int）。
- NONE：表示不需要缓存。

通过StorageLevel类的构造传参的方式进行控制，结构如下：

```
class StorageLevel private(useDisk : Boolean ,useMemory : Boolean ,deserialized : Boolean ,replication: Ini)
```

8. 提交方式

提交方式有3种：

- spark-submit（官方推荐）。
- sbt run。
- ava-jar。

提交时可以指定各种参数，例如：

```
./bin/spark-submit
-- class  <main- class >
--master <master-url>
--deploy-mode <deploy-mode>
--conf <key> = <value>
...  #  other options
```

```
<application-jar>
[application-arguments]
```

spark-submit 提交方式如下：

```
#Run application locally on 8 cores
./bin/spark-submit \
--class org.apache.spark.examples.SparkPi \
--master local[8]\
/path/to/examples.jar \
100
#Run on a Spark Standalone cluster in client deploy mode
./bin/spark-submit \
--class org.apache.spark.examples.SparkPi \
--master spark://207.184.161.138:7077
 --executor-memory 20G \
--total-executor-cores 100 \
/path/to/examples.jar \
1000
```

9. 监控

每个驱动程序都有一个 Web UI，通常位于端口 4040 上，用于显示与正在运行的任务、执行器和存储使用情况有关的信息。只需在 Web 浏览器中访问 http://<driver node>:4040 即可访问此 UI。

2.2 Linux 环境搭建

2.2.1 VirtualBox 虚拟机安装

本书将使用 VirtualBox 作为虚拟环境来安装 Linux 和 Hadoop。VirtualBox 最早由 SUN 公司开发。由于 SUN 公司目前已经被 Oracle 收购，因此可以在 Oracle 公司的官方网站上下载 VirtualBox 虚拟机软件的安装程序，产品地址为 https://www.virtualbox.org。到笔者写作本书时，VirtualBox 的最新版本为 7.0.6。

首先，到 VritualBox 的官方网站下载 Windows hosts 版本的 VirtualBox。下载地址为 https://www.virtualbox.org/wiki/Downloads，页面如图 2-4 所示。

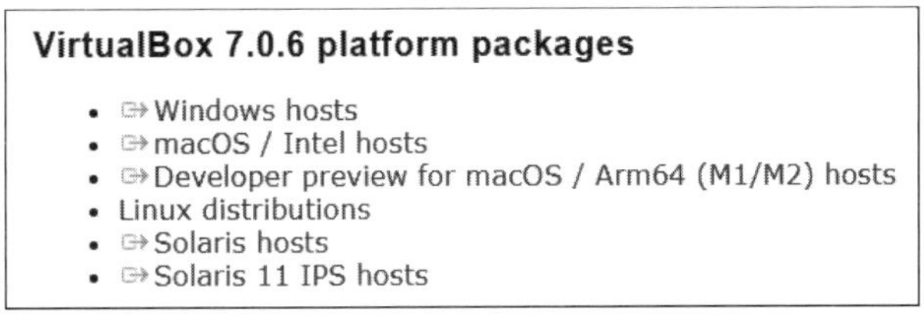

图 2-4

同时，VitualBox 需要虚拟化 CPU 的支持。如果安装的操作系统是不支持 x64 位的 CentOS，那么可以在宿主机开机时按 F12 键进入 BIOS 设置界面，并打开 CPU 的虚拟化设置界面。CPU 的虚拟化设置界面如图 2-5 所示。

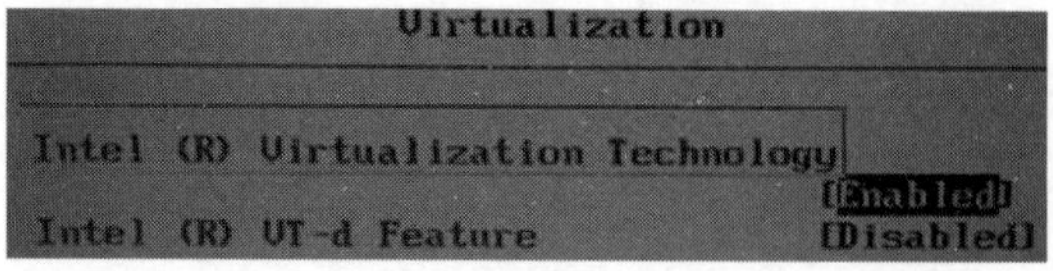

图 2-5

读者下载完成 VirtualBox 虚拟机后，自行安装即可。虚拟机的安装相对比较简单，以下是重要安装环节的截图。

网络功能的安装界面如图 2-6 所示，请单击"是"按钮。

图 2-6

网络功能下一步的安装界面如图 2-7 所示，请单击"安装"按钮。

图 2-7

网络功能安装成功后，会在"网络连接"里面多出一个名为 Virtual Box Host Only 的本地网卡，此网卡用于宿主机与虚拟机的通信，如图 2-8 所示。

图 2-8

2.2.2 安装 Linux 操作系统

本书将使用 CentOS 7 作为操作系统环境来学习和安装 Hadoop。首先需要下载 CentOS 操作系统，下载 Minimal（最小）版本的即可，因为我们使用 CentOS 时并不需要可视化界面。

CentOS 的官方网址为 https://www.centos.org/。CentOS 7 的下载页面如图 2-9 所示。

CentOS-7-x86_64-DVD-2009.iso	4.4 GiB	2020-11-04 19:37
CentOS-7-x86_64-DVD-2009.torrent	176.1 KiB	2020-11-06 22:44
CentOS-7-x86_64-Everything-2009.iso	9.5 GiB	2020-11-02 23:18
CentOS-7-x86_64-Everything-2009.torrent	380.6 KiB	2020-11-06 22:44
CentOS-7-x86_64-Minimal-2009.iso	973.0 MiB	2020-11-03 22:55
CentOS-7-x86_64-Minimal-2009.torrent	38.6 KiB	2020-11-06 22:44

图 2-9

下载完成以后，将得到一个 centOS-7-x86_64minimal-2009.iso 文件。注意文件名中的 2009 不是指 2009 年，而是指 2020 年 09 月发布的版本。接下来，启动 VirtualBox，开始安装 Linux 操作系统。

步骤01 在 VirtualBox 主界面的菜单栏上单击"新建"按钮，如图 2-10 所示。

图 2-10

步骤02 在 Name 字段中输入虚拟机的名称 CentOS7-201，保持 Folder 为默认值，选择操作系统镜像文件的路径，如图 2-11 所示。

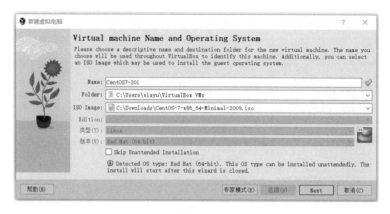

图 2-11

步骤 03 为新的系统分配内存，建议 4GB（最少 2GB）或以上，这要根据宿主机的内存而定。同时建议设置 CPU 为 2 颗。如图 2-12 所示。

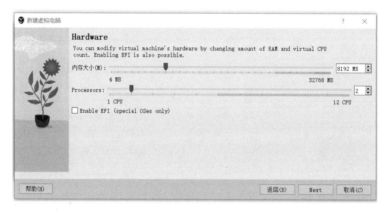

图 2-12

步骤 04 为新的系统创建虚拟硬盘，设置为动态增加，建议最大设置为 30GB 或以上。同时选择虚拟文件保存目录，默认情况下，会将虚拟化文件保存到 C:\盘当前用户的主目录上。笔者以为最好保存到非系统盘上，如 D:\OS 目录将是不错的选择。如图 2-13 所示。

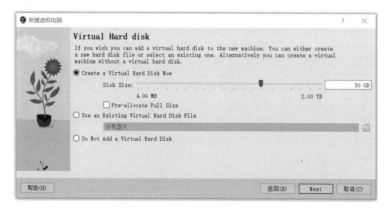

图 2-13

步骤 05 单击 Next 按钮，进入"摘要"界面，如图 2-14 所示。在界面上单击 Finish 按钮，关闭"新建虚拟电脑"界面，回到 VirtualBox 主界面，界面左侧栏已经显示我们新建的虚拟机 CentOS7-201，如图 2-15 所示。

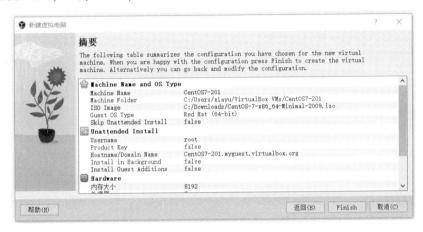

图 2-14

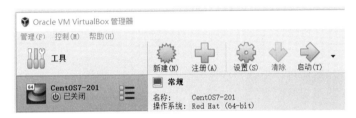

图 2-15

步骤 06 在如图 2-15 所示的 VirtualBox 主界面左侧选中 CentOS7-201 虚拟机，并单击右上方的"设置"按钮，打开"CentOS7-201 -设置"界面，如图 2-16 所示。

图 2-16

在"CentOS7-201 - 设置"界面左侧选择"网络",右侧会显示"网络"设置界面,将网卡 1 的连接方式设置为 NAT 用于连接外网,将网卡 2 的连接方式设置为 Host-Only 用于与宿主机进行通信。如果没有网卡 2,则需要关闭 Linux 虚拟机,在这个设置界面上对网卡 2 进行"启用网络链接"设置,并选择连接方式为"仅主机(Host-Only)网络"。

网卡 1 的设置如图 2-17 所示。

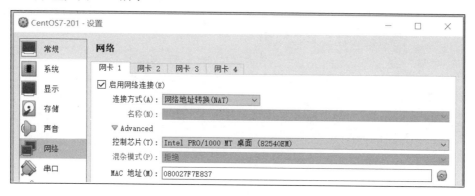

图 2-17

网卡 2 的设置如图 2-18 所示。

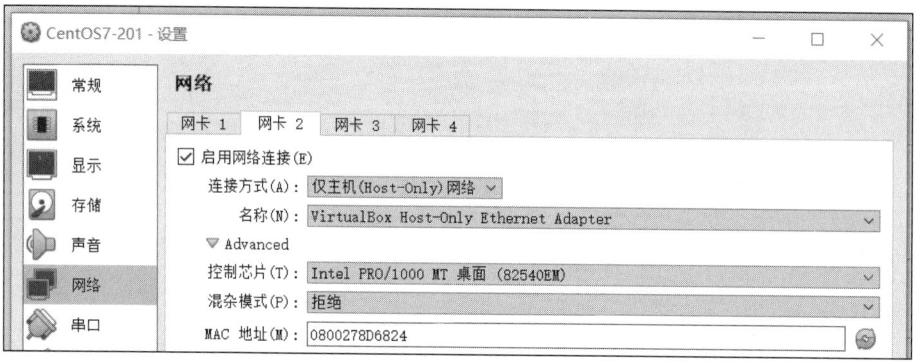

图 2-18

步骤 07 现在启动这个虚拟机,将会进入安装 CentOS 7 的界面,选择 Install CentOS Linux 7,接下来就开始安装 CentOS Linux 了,如图 2-19 所示。

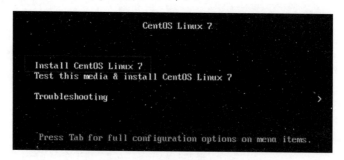

图 2-19

步骤 08 在安装过程中出现选择语言项目时,可以选择"中文"。选择安装介质,并进入安装位置,选择整个磁盘即可，如图 2-20、图 2-21 所示。注意，必须同时选择打开 CentOS 的网络，如图 2-22、图 2-23 所示，否则安装成功以后，CentOS 将没有网卡设置的选项。

图 2-20　　　　　　　　　　　　　　图 2-21

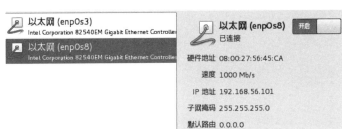

图 2-22　　　　　　　　　　　　　　图 2-23

步骤 09 在安装过程中，创建一个非 root 用户，并将它设置为管理员，如图 2-24、图 2-25 所示。在之后的操作中，笔者不建议使用 root 账户进行具体的操作，一般情况下，使用这个非 root 用户执行 sudo 即可以用 root 账户执行相关命令，输入的密码务必牢记。

图 2-24

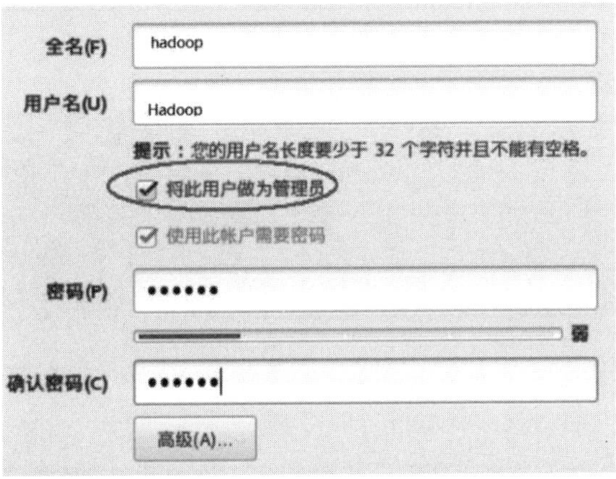

图 2-25

步骤 10 安装完成以后，重新启动，并测试是否可以使用之前创建的用户名和密码登录。刚开始安装完成后，请选择正常启动（即以有界面的方式启动），等我们设置好一些信息后，就可以选择无界面启动。

启动方式选择正常启动，如图 2-26 所示。

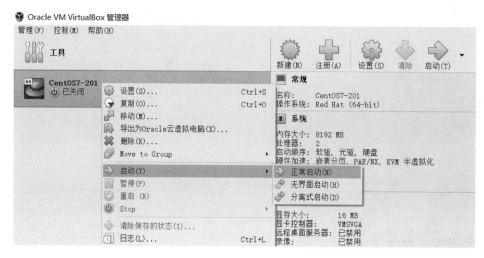

图 2-26

步骤 11 设置静态 IP 地址。

启动后，将显示如图 2-27 所示的界面，此时可以选择以 root 用户名和密码登录。注意，输入密码时将不会有任何的响应，不用担心，只要确认输入正确，按回车键后即可看到登录成功后的界面，如图 2-28 所示。

图 2-27

图 2-28

对于 Linux 系统来说，如果当前是 root 用户，将会显示"#"，如图 2-28 所示，root 用户登录成功后，将会显示"[root@server201 ~]#"，其中#表示当前为 root 用户。如果是非 root 用户，将显示为$。

设置静态 IP 地址，使用 vim 修改/etc/sysconfig/network-scripts/ifcfg-enp0s8，修改内容如下：

```
TYPE=Ethernet
```

```
PROXY_METHOD=none
BROWSER_ONLY=no
BOOTPROTO=**static**
DEFROUTE=yes
IPV4_FAILURE_FATAL=no
IPV6INIT=yes
IPV6_AUTOCONF=yes
IPV6_DEFROUTE=yes
IPV6_FAILURE_FATAL=no
IPV6_ADDR_GEN_MODE=stable-privacy
NAME=**enp0s8**
UUID=620377da-1744-4268-b6d6-a519d27e01c6
DEVICE=**enp0s8**
ONBOOT=yes
IPADDR=**192.168.56.201**
```

其中 IPADDR=192.168.56.201 为本 Linux 的 Host-Only 网卡地址，用于主机通信。输出完成以后，按 ESC 键，然后再输入":wq"保存退出即可。这是 vim 的基本操作，不了解的读者可以去网上查看一下 vim 的基本使用方法。

请牢记上面设置的 IP 地址，这个 192.168.56.201 的 IP 地址在后面会经常出现。现在我们可以关闭系统，并以非界面方式重新启动 CentOS。以后我们将使用 SSH 客户端登录此 CentOS。

上述文件是在配置了 Host-Only 网卡的情况下，才会存在 ifcfg-enp0s8。如果没有这个文件，请关闭 Linux，并重新添加 Host-Only 网卡，再进行配置。如果添加了 Host-Only 网卡后依然没有此文件，那么可以在相同目录下，复制 ifcfg-enp0s3 为 ifcfg-enp0s8 后再进行配置。

现在关闭 CentOS，以无界面方式启动，如图 2-29 所示。

图 2-29

注意：

（1）本书重点不是讲 VirtualBox 虚拟机的使用，因此这里只给出关键的操作步骤。

（2）在安装过程中，鼠标会在虚拟机和宿主机之间切换。如果要从虚拟机中退出鼠标，按键盘右边的 Ctrl 键即可。

（3）登录 Linux 系统后，随手执行命令 yum -y install vim 安装上 vim，方便使用。

（4）关于 Linux 命令请读者自行参考 Linux 手册，如：vim/vi、sudo、ls、cp、mv、tar、chmod、chown、scp、ssh-keygen、ssh-copy-id、cat、mkdir 等，它们将是后面经常使用的命令。

2.2.3　SSH 工具与使用

Linux 安装成功后，系统将自动运行 SSH 服务，读者可以选择 Xshell、CRT、MobaXterm 等客户端作为 Linux 远程命令行操作工具，同时配合它们的 xFtp 可以实现文件的上传与下载。Xshell 和 CRT 是收费软件，不过读者在安装时选择 free for school（学校免费版本）即可免费使用。

MobaXterm 个人版是免费的，本书选用它作为远程命令行执行、文件上传下载以及配置文件的编辑工具。到官网下载 MobaXterm 并安装完成以后，配置一下 SSH 即可登录 Linux 系统。配置很简单，在 MobaXterm 主界面上单击左上方的 Session 按钮，即可创建 SSH 连接，如图 2-30 所示。

图 2-30

单击 Session 按钮后，出现如图 2-31 所示的窗口，在窗口上单击 SSH 按钮，在相应的文本框中输入主机名称和登录用户名，再单击窗口下方的 OK 按钮保存一下。

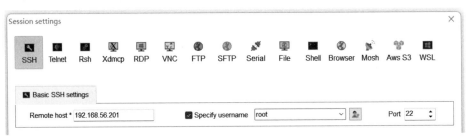

图 2-31

这时会打开 Linux 交互界面，提示输入 root 密码，输入密码不会有任何的回显，只要输入正确，按回车键即可登录，如图 2-32 所示。

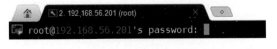

图 2-32　输入密码

root 用户登录成功以后的界面如图 2-33 所示，用户可以通过这个界面操作 Linux 系统。

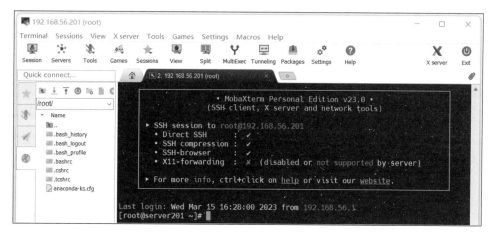

图 2-33

提示：使用 MobaXterm 工具连接了 CentOS 虚拟机，就不需要在虚拟机和宿主机之间来回切换。另外，还可以发起好几个连接访问 CentOS 虚拟机，使用起来非常方便。

还可以配置 SFTP 连接，方便本地下载的 Linux 软件包上传到 Linux 系统进行安装配置，Linux 系统上的配置文件也可以在本地编辑后自动上传。SFTP 登录界面如图 2-34 所示。

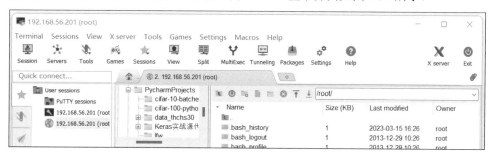

图 2-34

2.2.4　Linux 统一设置

后面配置 Hadoop 环境时将使用一些 Linux 统一的设置，在此一并列出。由于本次登录（见图 1-29）使用的是 root 账户，因此可以直接操作某些命令，不用添加 sudo 命令。

1. 配置主机名称

笔者的习惯是将"server+IP 地址最后一部分数字"作为主机名称，因此主机名设置为 server201，因为本主机设置的 IP 地址是 192.168.56.201。

```
# hostnamectl set-hostname server201
```

2. 修改 hosts 文件

在 hosts 文件最后添加的以下配置，可通过 vim /etc/hosts 命令进行修改。

```
192.168.56.201    server201
```

3. 关闭且禁用防火墙

```
# systemctl stop firewalld
# systemctl disable firewalld
```

4. 禁用 SElinux，需要重新启动

```
#vim /etc/selinux/config
SELINUX=disabled
```

5. 设置时间同步（可选）

```
#vim /etc/chrony.conf
```

删除所有的 server 配置，只添加：

```
server ntp1.aliyun.com iburst
```

重新启动 chronyd：

```
#systemctl restart chronyd
```

查看状态：

```
#chronyc sources -v
^* 120.25.115.20
```

如果结果显示"*"，则表示时间同步成功。

6. 在/usr/java 目录下安装 JDK1.8.x

usr 目录的意思是 unix system resource 目录，可以将 JDK1.8_x64 安装到此目录下。

首先去 Oracle 网站下载 JDK1.8 的 Linux 压缩包版本，页面如图 2-35 所示。

| Linux x64 Compressed Archive | 137.06 MB | jdk-8u281-linux-x64.tar.gz |

图 2-35

将压缩包上传到 Linux 并解压：

```
# mkdir /usr/java
# tar -zxvf jdk-8u361-linux-x64.tar.gz -C /usr/java/
```

7. 配置 JAVA_HOME 环境变量

```
# vim /etc/profile
```

在 profile 文件最后添加以下配置：

```
export JAVA_HOME=/usr/java/jdk1.8.0_361
export PATH=.:$PATH:$JAVA_HOME/bin
```

让环境变量生效：

```
# source /etc/profile
```

检查 Java 版本：

```
[root@localhost bin]# java -version
java version "1.8.0_361"
Java(TM) SE Runtime Environment (build 1.8.0_192-b12)
Java HotSpot(TM) 64-Bit Server VM (build 25.192-b12, mixed mode)
```

到此，基本的 Linux 运行环境就已经配置完成了。

提示：在 VirtualBox 虚拟机中，可以通过复制的方式为本小节已经做了统一设置的 CentOS 镜像文件创建副本，用于备份或者搭建集群。

2.3 Hadoop 安装与配置

2.3.1 Hadoop 安装环境准备

注意：读者可以从 2.2 节已配置好的 CentOS 虚拟机中复制一份干净的系统出来，用于本节搭建 Hadoop 运行环境，比如复制出来的 CentOS 虚拟机名字为 CenOS7-201。

本节讲解 Hadoop 伪分布式和完全分布式环境的搭建，以便提供 HDFS 与 YARN 功能与 Spark 集成。

Hadoop 伪分布式是在单机模式下运行 Hadoop，但用不同的 Java 进程模仿分布式运行中的各类节点。这种模式下，我们需要运行 5 个守护进程：

（1）3 个负责 HDFS 存储的进程，如图 2-36 所示。

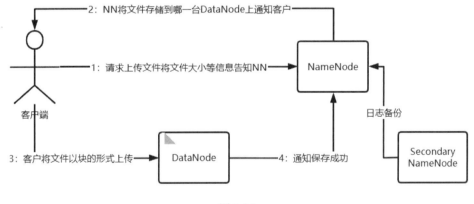

图 2-36

- NameNode 进程：作为主节点，主要负责分配数据存储的具体位置。
- SecondaryNameNode 进程：作为 NameNode 日志备份和恢复进程，用于避免数据丢失。
- DataNode 进程：作为数据的存储节点，接收客户端的数据读写请求。

（2）2 个负责 MapReduce 计算的进程：

- ResourceManager 进程：负责分配计算任务由哪一台主机执行。
- NodeManager 进程：负责执行计算任务。

在真实集群环境下，这些进程部署的一般规则是：

- 由于 NodeManger 需要读取 DataNode 上的数据用于执行计算，因此一般 DataNode 与 NodeManger 并存。
- 由于 NameNode 在运行时需要在内存中大量缓存文件块的数据，因此 NameNode 节点应该部署到内存比较大的主机上。
- 在真实的集群环境下，一般部署多个 NameNode 节点，互为备份和切换关系，且不再部署 SecondaryNameNode 进程。

在安装 Hadoop 环境之前，笔者有以下几点建议：

- 配置静态 IP 地址。虽然是单机模式，但也建议配置静态的 IP 地址，这有助于以后配置集群环境时固定 IP，养成良好的习惯。
- 修改主机名称为一个便于记忆的名称，如 server201，修改规则一般为将本机的 IP 地址最后一段作为服务器的后缀，如 IP 地址为 192.168.56.201，则可以修改本主机的名称为 server201。
- 由于启动 Hadoop 的各个进程使用的是 SSH，因此必须配置本机免密码登录。本章后面的步骤会讲到如何配置 SSH 免密码登录。配置 SSH 免密码登录的规则是在启动集群的主机上，向其他主机配置 SSH 免密登录，以便于操作机可以在不登录其他主机的情况下启动所需要的进程。
- 关闭防火墙。如果我们的 CentOS 7 没有安装防火墙，那么可以不用关闭了。如果已经安装了防火墙，则检查防火墙的状态，如果是运行状态请关闭防火墙并禁用防火墙。注意，在生产环境下，不要直接禁用防火墙，而是指定 Hadoop 的某些端口开放。
- 使用非 root 用户，前面章节我们创建了一个名为 hadoop 的用户，此用户同时属于 wheel 组（拥有此组的用户可以使用 sudo 命令，执行一些 root 用户的操作）。我们就以此用户作为执行命令的用户。

1. 配置静态 IP 地址

2.2.2 节已经讲解了静态 IP 的设置，此处再做一下补充。首先使用 SSH 登录 CentOS 7，然后使用 ifconfig 查看 IP 地址（如果没有 ifconfig 命令，可以使用 sudo yum -y install net-tools 安装 ifconfig 命令。其实在 CentOS 7 中，已经使用 ip addr 命令显示当前主机的 IP 地址，所以也可以不安装 net-tools）：

```
$ ifconfig
enp0s3: flags=4163<UP,BROADCAST,RUNNING,MULTICAST>  mtu 1500
    inet 10.0.2.15  netmask 255.255.255.0  broadcast 10.0.2.255
enp0s8: flags=4163<UP,BROADCAST,RUNNING,MULTICAST>  mtu 1500
    inet 192.168.56.201  netmask 255.255.255.0
```

上例显示为两个网卡，其中 enp0s3 的 IP 地址为 10.0.2.15，此网卡为 NAT 网络，可用于上网。enp0s8 的 IP 地址为 192.168.56.201，此网卡为 Host-Only 网络，用于与宿主机进行通信。我们要修改的就是 enp0s8 这个网卡，将它的 IP 地址设置为固定 IP。

IP 设置保存在/etc/sysconfig/network-scripts/ifcfg-enp0s8 文件中。使用 cd 命令，切换到这个目录下。使用 ls 显示这个目录下的所有文件，我们可能只会发现 ifcfg-enp0s3 这个文件，现在使用 cp 命令将 ifcfg-enp0s3 复制一份为 ifcfg-enp0s8。由于 etc 目录不属于 hadoop 用户，因此操作时需要添加 sudo 前缀。

```
$ sudo cp ifcfg-enp0s3 ifcfg-enp0s8
```

使用 vim 命令修改为静态 IP 地址：

```
$ sudo vim ifcfg-enp0s8
```

将原来的 dhcp 修改成 static，即静态的 IP 地址，并设置具体的 IP 地址。其中，每一个网卡都应该具有唯一的 UUID，因此建议修改任意的一个值，以便于与之前 enp0s3 的 UUID 不同。部分修改内容如下：

```
BOOTPROTO="static"
NAME="enp0s8"
UUID="d2a8bd92-cf0d-4471-8967-3c8aee78d101"
DEVICE="enp0s8"
IPADDR="192.168.56.201"
```

现在重新启动网络：

```
$ sudo systemctl restart network.service
```

重新启动网络后，再次查看：

```
[hadoop@server201 ~]$ ifconfig
enp0s3: flags=4163<UP,BROADCAST,RUNNING,MULTICAST>  mtu 1500
        inet 10.0.2.15  netmask 255.255.255.0  broadcast 10.0.2.255
enp0s8: flags=4163<UP,BROADCAST,RUNNING,MULTICAST>  mtu 1500
        inet 192.168.56.201  netmask 255.255.255.0  broadcast 0x20<link>
lo: flags=73<UP,LOOPBACK,RUNNING>  mtu 65536
        inet 127.0.0.1  netmask 255.0.0.0
```

可以发现 IP 地址已经发生了变化。

2. 修改主机名称

使用 hostname 命令，检查当前主机的名称：

```
$ hostname
localhost
```

使用 hostnamectl 命令，修改主机的名称：

```
$ sudo hostnamectl set-hostname server201
```

3. 配置 hosts 文件

hosts 文件是本地 DNS 解析文件。配置此文件，可以根据主机名找到对应的 IP 地址。

使用 vim 命令打开这个文件，并在文件中追加以下配置：

```
$ sudo vim /etc/hosts
192.168.56.201server201
```

4. 关闭防火墙

默认情况下，CentOS 7 没有安装防火墙。我们可以通过命令 sudo firewall-cmd --state 检查防火墙的状态，如果显示 command not found，则表示没有安装防火墙，此步可以忽略。以下命令检查防火墙的状态：

```
$ sudo firewall-cmd --state
running
```

running 表示防火墙正在运行。以下命令用于停止和禁用防火墙：

```
$ sudo systemctl stop firewalld.service
$ sudo systemctl disable firewalld.service
```

5. 配置免密码登录

配置免密码登录的主要目的就是在使用 hadoop 脚本启动 Hadoop 的守护进程时，不需要再提示用户输入密码。SSH 免密码登录的主要实现机制就是在本地生成一个公钥，然后将公钥配置到需要被免密登录的主机上，登录时将自己持有的私钥与公钥进行匹配，如果匹配成功，则登录成功，否则登录失败。

可以使用 ssh-keygen 命令生成公钥和私钥文件，并将公钥文件拷贝到被 SSH 登录的主机上。以下是 ssh-keygen 命令，输入以后直接按两次 Enter 键，即可以生成公钥和私钥文件：

```
[hadoop@server201 ~]$ ssh-keygen -t rsa
Generating public/private rsa key pair.
Enter file in which to save the key (/home/hadoop/.ssh/id_rsa):
Created directory '/home/hadoop/.ssh'.
Enter passphrase (empty for no passphrase):
Enter same passphrase again:
Your identification has been saved in /home/hadoop/.ssh/id_rsa.
Your public key has been saved in /home/hadoop/.ssh/id_rsa.pub.
The key fingerprint is:
SHA256:IDI032gBEDXhFVE1l6oYca5P4fkfIZRywyhgJ4Id/I4 hadoop@server201
The key's randomart image is:
+---[RSA 2048]----+
|=*%+*+..o ..     |
|.=oO.+.o +.      |
| +.*+= *.        |
|   +ooo=..       |
|    o = +S .     |
|   E + = ..      |
|      o ..       |
|       . . .     |
|         ..      |
+----[SHA256]-----+
```

如上的提示信息所示，生成的公钥和私钥文件将被放到~/.ssh/目录下。其中 id_rsa 文件为私钥文件，rd_rsa.pub 为公钥文件。现在我们再使用 ssh-copy-id 命令将公钥文件发送到目标主机。由于是登录本机，因此直接输入本机的主机名即可：

```
[hadoop@server201 ~]$ ssh-copy-id server201
/usr/bin/ssh-copy-id: INFO: Source of key(s) to be installed:
"/home/hadoop/.ssh/id_rsa.pub"
The authenticity of host 'server201 (192.168.56.201)' can't be established.
ECDSA key fingerprint is SHA256:KqSRs/H1WxHrBF/tfM67PeiqqcRZuK4ooAr+xT5Z4OI.
ECDSA key fingerprint is MD5:05:04:dc:d4:ed:ed:68:1c:49:62:7f:1b:19:63:5d:8e.
Are you sure you want to continue connecting (yes/no)? yes  输入 yes
/usr/bin/ssh-copy-id: INFO: attempting to log in with the new key(s), to filter
out any that are already installed
/usr/bin/ssh-copy-id: INFO: 1 key(s) remain to be installed -- if you are prompted
now it is to install the new keys
```

输入密码后按 Enter 键，将会提示成功信息：

```
hadoop@server201's password:
Number of key(s) added: 1
Now try logging into the machine, with:   "ssh 'server201'"
and check to make sure that only the key(s) you wanted were added.
```

此命令执行以后，会在~/.ssh 目录下多出一个用于认证的文件，其中保存了某个主机可以登录的公钥信息，这个文件为：~/.ssh/authorized_keys。如果读者感兴趣，可以使用 cat 命令查看这个文件中的内容，此文件中的内容就是 id_rsa.pub 文件中的内容。

现在再使用 ssh server201 登录本机，我们将会发现，不用输入密码即可直接登录成功。

```
[hadoop@server201 ~]$ ssh server201
Last login: Tue Mar  9 20:52:56 2021 from 192.168.56.1
```

2.3.2　Hadoop 伪分布式安装

经过上面环境的设置，我们已经可以正式安装 Hadoop 伪分布式了。在安装之前，先确定已经安装了 JDK1.8，并正确配置了 JAVA_HOME、PATH 环境变量。接下来创建一个工作目录/app，方便我们以 hadoop 账户安装、配置与运行 Hadoop 相关程序。

在磁盘根目录（/）下创建一个 app 目录，并授权给 hadoop 用户。我们会将 Hadoop 安装到此目录下。

步骤01 先切换到根目录下：

```
[hadoop@server201 ~]$ cd /
```

步骤02 添加 sudo 前缀，使用 mkdir 创建/app 目录：

```
[hadoop@server201 /]$ sudo mkdir /app
[sudo] hadoop 的密码：
```

步骤03 将此目录的所有权授予 hadoop 用户和 hadoop 组：

```
[hadoop@server201 /]$ sudo chown hadoop:hadoop /app
```

步骤04 su hadoop 账户，切换进入/app 目录：

```
[hadoop@server201 /]$ cd /app/
```

步骤 05 使用 ll -d 命令查看本目录的详细信息，可见此目录已经属于 hadoop 用户：

```
[hadoop@server201 app]$ ll -d
drwxr-xr-x 2 hadoop hadoop 6 3月   9 21:35 .
```

步骤 06 将 Hadoop 压缩包上传到 /app 目录下，并解压。

步骤 07 使用 ll 命令查看本目录：

```
[hadoop@server201 app]$ ll
总用量 386184
-rw-rw-r-- 1 hadoop hadoop 395448622 3月   9 21:40 hadoop-3.2.3.tar.gz
```

已经存在 hadoop-3.2.3.tar.gz 文件。

步骤 08 使用 tar -zxvf 命令解压此文件：

```
[hadoop@server201 app]$ tar -zxvf hadoop-3.2.3.tar.gz
```

步骤 09 查看 /app 目录，已经多出 hadoop-3.2.3 目录：

```
[hadoop@server201 app]$ ll
总用量 386184
drwxr-xr-x 9 hadoop hadoop       149 1月   3 18:11 hadoop-3.2.3
-rw-rw-r-- 1 hadoop hadoop 395448622 3月   9 21:40 hadoop-3.2.3.tar.gz
```

步骤 10 删除 hadoop-3.2.3.tar.gz 文件（此文件已不再被需要）：

```
[hadoop@server201 app]$ rm -rf hadoop-3.2.3.tar.gz
```

步骤 11 下面开始配置 Hadoop。Hadoop 的所有配置文件都在 hadoop-3.2.3/etc/hadoop 目录下，首先切换到此目录下，然后开始配置。

```
[hadoop@server201 hadoop-3.2.3]$ cd /app/hadoop-3.2.3/etc/hadoop/
```

在 Hadoop 的官网上有关于伪分布式配置的完整教程，它的网址是：

```
https://hadoop.apache.org/docs/stable/hadoop-project-dist/hadoop-common/SingleCluster.html#Configuration
```

大家也可以根据此教程进行 Hadoop 伪分布式的配置学习。

1. 配置 hadoop-env.sh 文件

hadoop-env.sh 文件是 Hadoop 的环境文件，在此文件中需要配置 JAVA_HOME 变量。在此文件的第 55 行，输入以下配置，然后按 ESC 键再输入 ":wq" 保存退出即可。

```
export JAVA_HOME=/usr/java/jdk1.8.0_281
```

2. 配置 core-site.xml 文件

core-site.xml 文件为 HDFS 的核心配置文件，用于配置 HDFS 的协议、端口号和地址。

注意：Hadoop 3.0 以后 HDFS 的端口号建议为 8020，但如果查看 Hadoop 官网的示例，依然使用的是 Hadoop 2 之前的端口 9000，以下配置笔者将使用 8020 端口，只要保证配置的端口没有被占用即可。配置时，注意字母大小写。

使用 vim 打开 core-site.xml 文件，进入编辑模式：

```
[hadoop@server201 hadoop]$ vim core-site.xml
```

在\<configuration\>\</configuration\>两个标签之间输入以下内容：

```xml
<property>
    <name>fs.defaultFS</name>
    <value>hdfs://server201:8020</value>
</property>
<property>
    <name>hadoop.tmp.dir</name>
    <value>/app/datas/hadoop</value>
</property>
```

配置说明如下：

（1）fs.defaultFS 用于配置 HDFS 的主协议，默认为 file:///。

（2）hadoop.tmp.dir 用于指定 NameNode 日志及数据的存储目录，默认为/tmp。需要使用 hadoop 账户在/app 目录下依次创建 datas、hadoop 目录，与此配置对应。

3. 配置 hdfs-site.xml 文件

hdfs-site.xml 文件用于配置 HDFS 的存储信息。使用 vim 打开 hdfs-site.xml 文件，并在\<configuration\>\</configuration\>标签中输入以下内容：

```xml
<property>
    <name>dfs.namenode.name.dir</name>
    <value>/app/hadoop-3.2.3/dfs/name</value>
</property>
<property>
    <name>dfs.datanode.data.dir</name>
    <value>/app/hadoop-3.2.3/dfs/data</value>
</property>
<property>
    <name>dfs.replication</name>
    <value>1</value>
</property>
<property>
    <name>dfs.permissions.enabled</name>
    <value>false</value>
</property>
```

配置说明如下：

（1）dfs.replication 用于指定文件块的副本数量。HDFS 特别适合于存储大文件，它会将大文件切分成每 128MB 一块，存储到不同的 DataNode 节点上，且默认会为每一块备份 2 份，共 3 份，即此配置的默认值为 3，最大为 512。由于我们只有一个 DataNode，因此这里将文件副本数量修改为 1。

（2）dfs.permissions.enabled 用于指定访问时是否检查安全，默认为 true，为了方便访问，暂时修改为 false。

4. 配置 mapred-site.xml 文件

通过名称可见，此文件是用于配置 MapReduce 的配置文件。使用 vim 打开此文件，并在 <configuration> 标签中输入以下配置信息：

```
<property>
    <name>mapreduce.framework.name</name>
    <value>yarn</value>
</property>
```

配置说明如下：

mapreduce.framework.name 用于指定调用方式，这里指定使用 YARN 作为任务调用方式。

5. 配置 yarn-site.xml 文件

由于上面指定使用 YARN 作为任务调用公式，因此这里需要配置 YARN 的配置信息，同样使用 vim 编辑 yarn-site.xml 文件，并在<configuration>标签中输入以下内容：

```
<property>
    <name>yarn.resourcemanager.hostname</name>
    <value>server201</value>
</property>
<property>
    <name>yarn.nodemanager.aux-services</name>
    <value>mapreduce_shuffle</value>
</property>
```

通过 hadoop classpath 命令获取所有 classpath 的目录，然后配置到上述文件中。

由于还没有配置 Hadoop 的环境变量，因此需要输入完整的 Hadoop 运行路径：

```
[hadoop@server201 hadoop]$ /app/hadoop-3.2.3/bin/hadoop classpath
```

命令完成后，将显示所有 classpath 信息：

```
/home/hadoop/program/hadoop-3.2.3/etc/hadoop:/home/hadoop/program/hadoop-3.2.3/share/hadoop/common/lib/*:/home/hadoop/program/hadoop-3.2.3/share/hadoop/common/*:/home/hadoop/program/hadoop-3.2.3/share/hadoop/hdfs:/home/hadoop/program/hadoop-3.2.3/share/hadoop/hdfs/lib/*:/home/hadoop/program/hadoop-3.2.3/share/hadoop/hdfs/*:/home/hadoop/program/hadoop-3.2.3/share/hadoop/mapreduce/lib/*:/home/hadoop/program/hadoop-3.2.3/share/hadoop/mapreduce/*:/home/hadoop/program/hadoop-3.2.3/share/hadoop/yarn:/home/hadoop/program/hadoop-3.2.3/share/hadoop/yarn/lib/*:/home/hadoop/program/hadoop-3.2.3/share/hadoop/yarn/*
```

将上述信息复制一下，并使用 MobaXterm 的 XFTP 功能配置到 yarn-site.xml 文件中：

```
<property>
    <name>yarn.application.classpath</name>
<value>/home/hadoop/program/hadoop-3.2.3/etc/hadoop:/home/hadoop/program/hadoop-3.2.3/share/hadoop/common/lib/*:/home/hadoop/program/hadoop-3.2.3/share/hadoop/common/*:/home/hadoop/program/hadoop-3.2.3/share/hadoop/hdfs:/home/hadoop/program/hadoop-3.2.3/share/hadoop/hdfs/lib/*:/home/hadoop/program/hadoop-3.2.3/shar
```

```
e/hadoop/hdfs/*:/home/hadoop/program/hadoop-3.2.3/share/hadoop/mapreduce/lib/*:
/home/hadoop/program/hadoop-3.2.3/share/hadoop/mapreduce/*:/home/hadoop/program
/hadoop-3.2.3/share/hadoop/yarn:/home/hadoop/program/hadoop-3.2.3/share/hadoop/
yarn/lib/*:/home/hadoop/program/hadoop-3.2.3/share/hadoop/yarn/*</value>
    </property>
```

配置说明如下:

(1) yarn.resourcemanager.hostname 用于指定 ResourceManger 的运行主机,默认为 0.0.0.0,即本机。

(2) yarn.nodemanager.aux-services 用于指定执行计算的方式为 mapreduce_shuffle。

(3) yarn.application.classpath 用于指定运算时的类加载目录。注意:这个配置的<value>标签及其值需要在同一行上,否则运行 Hadoop 会出错。

6. 配置 workers 文件

workers 文件在以前的版本中叫作 slaves,它们所起的作用是一样的。workers 文件主要用于在启动 Hadoop 时启动 DataNode 和 NodeManager。

编辑 workers 文件,并输入本地主机名称:

```
server201
```

7. 配置 Hadoop 环境变量

编辑/etc/profile 文件:

```
$ sudo vim /etc/profile
```

在文件里添加以下内容:

```
export HADOOP_HOME=/app/hadoop-3.2.3
export PATH=$PATH:$HADOOP_HOME/bin
```

使用 source 命令让环境变量生效:

```
$ source /etc/profile
```

使用 hdfs version 命令查看环境变量是否生效,如果配置成功,则会显示 Hadoop 的版本:

```
[hadoop@server201 hadoop]$ hdfs version
Hadoop 3.2.3
Source code repository Unknown -r 7a3bc90b05f257c8ace2f76d74264906f0f7a932
Compiled by hexiaoqiao on 2021-01-03T09:26Z
Compiled with protoc 2.5.0
From source with checksum 5a8f564f46624254b27f6a33126ff4
This command was run using
/app/hadoop-3.2.3/share/hadoop/common/hadoop-common-3.2.3.jar
```

8. 初始化 Hadoop 的文件系统

在使用 Hadoop 之前,必须先初始化 HDFS 文件系统,初始化的文件系统将会生成在 hadoop.tmp.dir 配置的目录下,即上面配置的/app/datas/hadoop 目录下:

```
$ hdfs namenode -format
```

在执行命令完成以后,若在输出的日志中找到以下这句话,即可确认初始化成功:

```
Storage directory /opt/hadoop_tmp_dir/dfs/name has been successfully formatted.
```

9. 启动 HDFS 和 YARN

启动和停止 HDFS 及 YARN 的脚本在$HADOOP_HOME/sbin 目录下,其中 start-dfs.sh 为启动 HDFS 的脚本,start-yarn.sh 为启动 ResourceManager 的脚本。以下代码分别启动 HDFS 和 YARN:

```
[hadoop@server201 /]$ /app/hadoop-3.2.3/sbin/start-dfs.sh
[hadoop@server201 /]$ /app/hadoop-3.2.3/sbin/start-yarn.sh
```

启动完成以后,通过 jps 命令来查看 Java 进程快照,我们会发现有 5 个进程正在运行:

```
[hadoop@server201 /]$ jps
12369 NodeManager
12247 ResourceManager
11704 NameNode
12025 SecondaryNameNode
12686 Jps
11839 DataNode
```

其中 NameNode、SecondaryNameNode、DataNode 是通过 start-dfs.sh 脚本启动的,ResourceManager 和 NodeManager 是通过 start-yarn.sh 脚本启动的。

启动 HDFS 和 YARN 成功以后,也可以通过 http://server201:9870(宿主机上使用 http://192.168.56.201:9870)查看 NameNode 的信息,如图 2-37 所示。

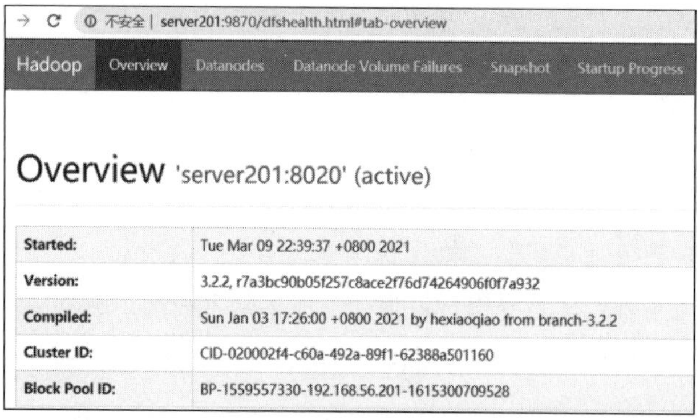

图 2-37

可以通过 http://server201:8088(宿主机上使用 http://192.168.56.201:8088)查看 MapReduce 的信息,如图 2-38 所示。

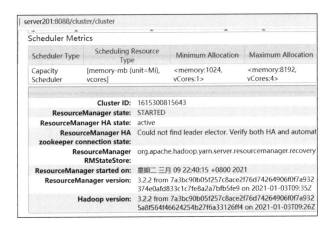

图 2-38

10. 关闭 HDFS 和 YARN

执行 stop-dfs.sh 和 stop-yarn.sh 关闭 HDFS 和 YARN：

```
[hadoop@server201 /]$ /app/hadoop-3.2.3/sbin/stop-yarn.sh
Stopping nodemanagers
Stopping resourcemanager
[hadoop@server201 /]$ /app/hadoop-3.2.3/sbin/stop-dfs.sh
Stopping namenodes on [server201]
Stopping datanodes
Stopping secondary namenodes [server201]
```

至此，Hadoop 单机即伪分布式运行模式安装并配置成功。但是万里长征，我们这才是小小的一步。以下将继续学习 Hadoop 完全分布式环境的搭建。

2.3.3　Hadoop 完全分布式环境搭建

在安装 Spark 之前，我们先把 Hadoop 完全分布式（集群）环境搭建起来。在 Hadoop 的集群中，有一个 NameNode，一个 ResourceManager；在高可靠的集群环境中，可以拥有两个 NameNode 和两个 ResourceManager；在 Hadoop3 以后，同一个 NameService 可以拥有 3 个 NameNode。由于 NameNode 和 ResourceManager 是两个主要的服务，因此建议将它们部署到不同的服务器上。

下面我们以 3 台服务器为例，来快速学习 Hadoop 的完全分布式环境的安装，这对深入了解后面讲解的 Spark 集群运行的基本原理非常有用。以下将分步骤为读者详解如何搭建 Hadoop 的完全分布式集群。

注意：可以利用虚拟机软件 VirtualBox 复制出来的 CentOS 镜像文件，快速搭建 3 个 CentOS 虚拟主机来做集群。

完整的集群主机配置如表 2-2 所示。

表2-2 集群主机配置表

IP/主机名	虚拟机	进程	软件
192.168.56.101/server101	CentOS7-101 8G 内存，2 核	NameNode SecondaryNameNode ResourceManager DataNode NodeManager	JDK HADOOP
192.168.56.102/server102	CentOS7-102 2G+内存，1 核	DataNode NodeManager	JDK HADOOP
192.168.56.103/server103	CentOS7-103 2G+内存，1 核	DataNode NodeManager	JDK Hadoop

从表 2-2 中可以看出，server101 运行的进程比较多，且 NameNode 运行在上面，因此这台主机需要更多的内存。

这里推荐读者使用 VirtualBox 把 2.2.2 节配置好虚拟机 CentOS7-201 复制出来，稍微做些修改，即可快速搭建 Hadoop 完全分布式环境。修改如下：

（1）把 CentOS7-201 复制为 CentOS7-101，按下面的 步骤01~步骤03 核对和修改相关配置，已经配置好的可以跳过去。

（2）把 CentOS7-101 复制为 CentOS7-102、CentOS7-103，由于此时 CentOS7-101 已基本配置好了，因此复制出来的 CentOS7-102、CentOS7-103 只要改一下主机名称和 IP 地址即可。

（3）3 台虚拟机配置好了以后，再按下面的 步骤04~步骤05 运行这个完全分布式集群。

Hadoop 完全分布式环境如图 2-39 所示。

图 2-39 Hadoop 完全分布式环境

操作步骤如下：

步骤01 准备工作。

（1）所有主机安装 JDK1.8+。建议将 JDK 安装到不同主机的相同目录下，这样可以减少修改配置文件的次数。

（2）在主节点（即执行 start-dfs.sh 和 start-yarn.sh 的主机）上向所有其他主机做 SSH 免密码登录。

（3）修改所有主机的主机名称和 IP 地址。

（4）配置所有主机的 hosts 文件，添加主机名和 IP 地址的映射：

```
192.168.56.101 server101
192.168.56.102 server102
192.168.56.103 server103
```

（5）使用以下命令关闭所有主机上的防火墙：

```
systemctl stop firewalld
systemctl disable firewalld
```

步骤02 在 server101 上安装 Hadoop。

可以将 Hadoop 安装到任意的目录下，如在根目录下创建/app 然后授予 hadoop 用户即可。

将 hadoop-3.2.3.tar.gz 解压到/app 目录下，并配置/app 目录属于 hadoop 用户：

```
$ sudo tar -zxvf hadoop-3.2.3.tag.gz -C /app/
```

将/app 目录及子目录授权给 hadoop 用户和 hadoop 组：

```
$suto chown hadoop:hadoop -R /app
```

接下来的配置文件都在/app/hadoop-3.2.3/etc/hadoop 目录下。配置文件 hadoop-env.sh：

```
export JAVA_HOME=/usr/java/jdk1.8.0_361
```

配置文件 core-site.xml：

```xml
<configuration>
    <property>
        <name>fs.defaultFS</name>
        <value>hdfs://server101:8020</value>
    </property>
    <property>
        <name>hadoop.tmp.dir</name>
        <value>/app/datas/hadoop</value>
    </property>
</configuration>
```

配置文件 hdfs-site.xml：

```xml
<configuration>
    <property>
        <name>dfs.namenode.name.dir</name>
        <value>/app/hadoop-3.2.3/dfs/name</value>
    </property>
```

```
    <property>
        <name>dfs.datanode.data.dir</name>
        <value>/app/hadoop-3.2.3/dfs/data</value>
    </property>
    <property>
        <name>dfs.replication</name>
        <value>3</value>
    </property>
    <property>
        <name>dfs.permissions.enabled</name>
        <value>false</value>
    </property>
</configuration>
```

配置文件 mapred-site.xml：

```
<configuration>
    <property>
        <name>mapreduce.framework.name</name>
        <value>yarn</value>
    </property>
</configuration>
```

配置文件 yarn-site.xml：

```
<configuration>
    <property>
        <name>yarn.nodemanager.aux-services</name>
        <value>mapreduce_shuffle</value>
    </property>
    <property>
        <name>yarn.resourcemanager.hostname</name>
        <value>server101</value>
    </property>
    <property>
        <name>yarn.application.classpath</name>
        <value>请自行执行 hadoop classpath 命令并将结果填入</value>
    </property>
</configuration>
```

配置 workers 配置文件。workers 配置文件用于配置执行 DataNode 和 NodeManager 的节点：

```
server101
server102
server103
```

步骤03 使用 scp 将 Hadoop 分发到其他主机。

由于 scp 会在网络上传递文件，而 hadoop/share/doc 目录下都是文档，没有必要进行复制，因此可以删除这个目录。

删除 doc 目录：

```
$ rm -rf /app/hadoop-3.2.3/share/doc
```

然后复制 server101 的文件到其他两台主机的相同目录下：

```
$scp -r /app/hadoop-3.2.3   server102:/app/
$scp -r /app/hadoop-3.2.3   server103:/app/
```

步骤04 在 server101 上格式化 NameNode。

首先需要在 server101 上配置 Hadoop 的环境变量，打开/etc/profile 文件：

```
$ sudo vim /etc/profile
```

在文件最后追加：

```
export HADOOP_HOME=/app/hadoop-3.2.3
export PATH=$PATH:$HADOOP_HOME/bin
```

在 server101 上执行 namenode 初始化命令：

```
$ hdfs namenode -format
```

步骤05 启动 HDFS 和 YARN。

在 server101 上执行启动工作，由于配置了集群，因此该启动过程会以 SSH 方式登录其他两台主机，并分别启动 DataNode 和 NodeManager：

```
$ /app/hadoop-3.2.3/sbin/start-dfs.sh
$ /app/hadoop-3.2.3/sbin/start-yarn.sh
```

启动完成后，通过宿主机的浏览器查看 9870 端口，页面会显示集群情况。

访问 http://192.168.56.101:9870 地址，会发现 3 个 DataNode 节点同时存在，如图 2-40 所示。

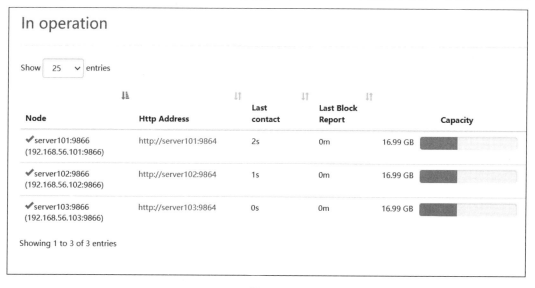

图 2-40

访问 http://192.168.56.101:8088 地址，会发现集群的 3 个活动节点同时存在，如图 2-41 所示。

Show 20 entries

Node Labels	Rack	Node State	Node Address	Node HTTP Address	Last health-update
	/default-rack	RUNNING	server101:36315	server101:8042	星期日 三月 19 12:08:57 +0800 2023
	/default-rack	RUNNING	server102:35227	server102:8042	星期日 三月 19 12:08:56 +0800 2023
	/default-rack	RUNNING	server103:37502	server103:8042	星期日 三月 19 12:08:55 +0800 2023

Showing 1 to 3 of 3 entries

图 2-41

步骤 06 执行 MapReduce 测试一下集群。

最后，建议执行 MapReduce 测试一下集群，比如执行以下 WordCount 示例，如果可以顺利执行完成，则说明整个集群的配置都是正确的。首先创建一个文本文件 a.txt，并输入几行英文句子：

```
[hadoop@server101 ~]$ vim a.txt
Hello This is
a Very Sample MapReduce
Example of Word Count
Hope You Run This Program Success!
```

分别执行以下命令：

```
[hadoop@server101 ~]$ hdfs dfs -mkdir -p /home/hadoop
[hadoop@server101 ~]$ hdfs dfs -mkdir /home/hadoop
[hadoop@server101 ~]$ hdfs dfs -put ./a.txt /home/hadoop
[hadoop@server101 ~]$ yarn jar
/app/hadoop-3.2.3/share/hadoop/mapreduce/hadoop-mapreduce-examples-3.2.3.jar
wordcount ~/a.txt /out002
```

2.4 Spark 安装与配置

Spark 的运行不一定必须安装 Hadoop 环境，但是 Spark 本身没有用于分布式存储的任何文件系统。由于许多大数据项目需要处理 PB 级的数据，而这些数据需要存储在分布式存储系统中，因此，在这种情况下，Spark 需要与 Hadoop 的分布式文件系统（HDFS）、资源管理器 YARN 或者其他第三方存储系统一起使用。本节主要讲解以下 4 种 Spark 安装方式：

- 本地模式安装。
- 伪分布模式安装。
- 完全分布模式安装。

- Spark on YARN 安装。

无论哪一种安装方式，都需要 JDK1.8+环境，所以，需先准备好 JDK 环境，并正确配置 JAVA_HOME 和 PATH 环境变量。

本书环境特别说明：为了读者能够快速上手学习，本书第 3 章、第 4 章、第 6 章、第 7 章、第 8 章、第 10 章、第 11 章均采用 Spark 本地运行模式，本地模式的安装参见 2.4.1 节。使用本地运行模式的章节中涉及编程实现的部分章节一般不需要安装 Spark，只需要在项目中引入与 Spark 相关的 JAR 包即可，程序中则通过 SparkContext 的 setMaster("local[*]")方法进行设置。

2.4.1 本地模式安装

本地模式的安装比较简单，直接启动上一节安装配置好的 CentOS7-201 虚拟机，以 hadoop 账户登录 Linux，下载并解压 Spark 安装文件就可以运行。这种模式可以让我们快速了解 Spark。下面介绍一下 Spark 本地模式的安装。

步骤 01 下载 Spark 安装文件，解压并配置环境变量：

```
[hadoop@server201 app]$ wget 
https://archive.apache.org/dist/spark/spark-3.3.1/spark-3.3.1-bin-hadoop3.tgz
[hadoop@server201 app]$ tar -zxvf spark-3.3.1-bin-hadoop3.tgz -C /app/
[hadoop@server201 app]$ sudo vim /etc/profile
export SPARK_HOME=/app/spark-3.3.1
export PATH=$PATH:$SPARK_HOME/bin
[hadoop@server201 app]$ source /etc/profile
```

步骤 02 配置完成以后，先通过 spark-shell 查看帮助和版本信息，还可以使用--help 查看所有选项的帮助信息：

```
[hadoop@server201 app]$ spark-shell --help
Usage: ./bin/spark-shell [options]
Scala REPL options:
  -I <file>                   preload <file>, enforcing line-by-line
                              interpretation
Options:
  --master MASTER_URL         spark://host:port, mesos://host:port, yarn,
                              k8s://https://host:port, or local (Default:
                              local[*]).
...
```

步骤 03 查看 Spark 的版本，直接使用--version 参数即可：

```
[hadoop@server201 app]$ spark-shell --version
Spark Version 3.3.1
Using Scala version 2.13.8, Java HotSpot(TM) 64-Bit Server VM, 1.8.0_361
Branch HEAD
Compiled by user ubuntu on 2021-02-22T01:33:19Z
Revision 1d550c4e90275ab418b9161925049239227f3dc9
```

```
Url https://github.com/apache/spark
Type --help for more information.
```

其中显示 Spark 的版本为 3.3.1，Scala 的版本为 2.13.8。

步骤 04 使用 spark-shell 启动 Spark 客户端。

使用 spark-shell 启动 Spark 客户端，通过--master 指定为 local 模式，通过 local[2]指定使用两核：

```
$ spark-shell --master local[2]
Welcome to
      ____              __
     / __/__  ___ _____/ /__
    _\ \/ _ \/ _ `/ __/  '_/
   /___/ .__/\_,_/_/ /_/\_\   version 3.3.1
      /_/

Using Scala version 2.13.8 (Java HotSpot(TM) 64-Bit Server VM, Java 1.8.0_361)
Type in expressions to have them evaluated.
Type :help for more information.
scala>
```

下面我们运行官方提供的 WordCount 示例，示例中存在的一些方法读者可能尚不明白，不过没有关系，在后面的章节中我们会详细讲解。

首先通过 sc 获取 SparkContext 对象，加载一个文件到内存中：

```
scala> val file = sc.textFile("file:///app/hadoop-3.2.3/NOTICE.txt");
val file: org.apache.spark.rdd.RDD[String] =
file:///app/hadoop-3.2.3/NOTICE.txt MapPartitionsRDD[1] at textFile at <console>:1
```

然后使用一系列的算子对文件对象进行处理——先按空格键和 Enter 键进行分割，然后使用 map 将数据组合成(key,value)形式，最后使用 reduceByKey 算子将 key 进行合并：

```
scala> val words = file.flatMap(_.split("\\s+")).map((_,1)). reduceByKey(_+_);
val words: org.apache.spark.rdd.RDD[(String, Int)] = ShuffledRDD[4] at
reduceByKey at <console>:1
```

最后调用 collect 方法输出结果：

```
scala> words.collect
val res0: Array[(String, Int)] = Array((this,2), (is,1), (how,1), (into,2),
(something,1), (hive.,2), (file,1), (And,1), (process,1), (you,2), (about,1),
(wordcount,1), (import,1), (a,1), (text,1), (be,1), (to,2), (in,1), (tell,1),
(for,1), (must,1))
```

上例的运算过程，也可以打开宿主机浏览器访问 http://192.168.56.201:4040 查看 DAG 运行效果，如图 2-42 所示。

图 2-42

从图 2-42 中可以看出 reduceByKey 引发了第二个 Stage，从 Stage0 到 Stage1 将会引发 shuffle，这也是区分转换算子和行动算子的主要依据。

通过上面的示例可以看出，在本地模式下运行 Spark 不需要事先启动任何的进程；启动 spark-shell 后，可以通过 SparkContext 读取本地文件系统目录下的文件。

操作完成以后，输入:quit 即可退出：

```
scala> :quit
[hadoop@server201 app]$
```

2.4.2 伪分布模式安装

伪分布模式也是在一台主机上运行，我们直接使用上一小节配置好的 CentOS7-201 虚拟机。伪分布模式需要启动 Spark 的两个进程，分别是 Master 和 Worker。启动后，可通过 8080 端口查看 Spark 的运行状态。伪分布模式安装需要修改一个配置文件 SPARK_HOME/conf/workers，添加一个 worker 节点，然后通过 SPARK_HOME/sbin 目录下的 start-all.sh 启动 Spark 集群。完整的 Spark 伪分布模式安装的操作步骤如下：

步骤01 配置 SSH 免密码登录。

由于启动 Spark 需要远程启动 Worker 进程，所以需要配置从 start-all.sh 的主机到 worker 节点的 SSH 免密码登录（如果之前已经配置过此项，那么可以不用重复配置）：

```
$ ssh-keygen -t rsa
$ ssh-copy-id server201
```

步骤02 修改配置文件。

在 spark-env.sh 文件中添加 JAVA_HOME 环境变量（在最前面添加即可）：

```
$ vim /app/spark-3.3.1/sbin/spark-env.sh
export JAVA_HOME=/usr/java/jdk1.8.0-361
```

修改 workers 配置文件：

```
$ vim /app/spark-3.3.1/conf/workers
server201
```

步骤 03 执行 start-all.sh 启动 Spark。

使用 start-all.sh 启动 Spark：

```
$ /app/spark-3.3.1/sbin/start-all.sh
```

启动完成以后，会有两个进程，分别是 Master 和 Worker：

```
[hadoop@server201 sbin]$ jps
2128 Worker
2228 Jps
2044 Master
```

查看启动日志可知，可以通过访问 8080 端口查看 Web 界面：

```
$ cat /app/spark-3.3.1/logs/spark-hadoop-org.apache.spark.deploy.master.Master-1-server201.out
21/03/22 22:03:47 INFO Utils: Successfully started service 'sparkMaster' on port 7077.
21/03/22 22:03:47 INFO Master: Starting Spark master at spark://server201:7077
21/03/22 22:03:47 INFO Master: Running Spark version 3.3.1
21/03/22 22:03:47 INFO Utils: Successfully started service 'MasterUI' on port 8080.
21/03/22 22:03:47 INFO MasterWebUI: Bound MasterWebUI to 0.0.0.0, and started at http://server201:8080
21/03/22 22:03:47 INFO Master: I have been elected leader! New state: ALIVE
21/03/22 22:03:50 INFO Master: Registering worker 192.168.56.201:34907 with 2 cores, 2.7 GiB RAM
```

步骤 04 通过 netstat 命令查看端口使用情况，我们会发现一共有两个端口被占用，它们分别是 7077 和 8080。

```
[hadoop@server201 sbin]$ netstat -nap | grep java
(Not all processes could be identified, non-owned process info
 will not be shown, you would have to be root to see it all.)
    tcp6       0      0 :::8080                 :::*                    LISTEN      2044/java
    tcp6       0      0 :::8081                 :::*                    LISTEN      2128/java
    tcp6       0      0 192.168.56.201:34907    :::*                    LISTEN      2128/java
    tcp6       0      0 192.168.56.201:7077     :::*                    LISTEN      2044/java
    tcp6       0      0 192.168.56.201:53630    192.168.56.201:7077     ESTABLISHED 2128/java
    tcp6       0      0 192.168.56.201:7077     192.168.56.201:53630    ESTABLISHED
```

```
2044/java
    unix   2      [ ]          STREAM       CONNECTED      53247        2044/java
    unix   2      [ ]          STREAM       CONNECTED      55327        2128/java
    unix   2      [ ]          STREAM       CONNECTED      54703        2044/java
    unix   2      [ ]          STREAM       CONNECTED      54699        2128/java
[hadoop@server201 sbin]$ jps
2128 Worker
2243 Jps
2044 Master
```

步骤 05 查看 8080 端口。

在宿主机浏览器中直接输入 http://192.168.56.201:8080 查看 Spark 运行的状态,如图 2-43 所示。

图 2-43

步骤 06 测试集群是否运行。

依然是使用 spark-shell,通过 --master 指定 spark://server201:7077 的地址即可以使用这个集群:

```
$ spark-shell --master spark://server201:7077
```

然后我们可以再做一次 2.4.1 节的 WordCount 测试。

读取文件:

```
scala> val file = sc.textFile("file:///app/hadoop-3.2.3/NOTICE.txt");
file: org.apache.spark.rdd.RDD[String] = file:///app/hadoop-3.2.3/NOTICE.txt
MapPartitionsRDD[1] at textFile at <console>:1
```

根据空格键和 Enter 键将字符串分割为一个一个的单词:

```
scala> val words = file.flatMap(_.split("\\s+"));
words: org.apache.spark.rdd.RDD[String] = MapPartitionsRDD[2] at flatMap at
<console>:1
```

进行统计,每一个单词初始统计为 1:

```
scala> val kv = words.map((_,1));
kv: org.apache.spark.rdd.RDD[(String, Int)] = MapPartitionsRDD[3] at map at
<console>:1
```

根据 key 进行统计计算：

```
scala> val worldCount = kv.reduceByKey(_+_);
worldCount: org.apache.spark.rdd.RDD[(String, Int)] = ShuffledRDD[4] at
reduceByKey at <console>:1
```

最后，输出结果并在每一行中添加一个制表符：

```
scala> worldCount.collect.foreach(kv=>println(kv._1+"\t"+kv._2));
this      2
is        1
how       1
into      2
something     1
hive.     2
file      1
And       1
process   1
you       2
about     1
wordcount     1
```

如果在运行时查看后台进程，我们将会发现多出以下两个进程：

```
[hadoop@server201 ~]# jps
12897 Worker
13811 SparkSubmit
13896 CoarseGrainedExecutorBackend
12825 Master
14108 Jps
```

SparkSubmit 为一个客户端，与 Running Application 对应。CoarseGrainedExecutorBackend 用于接收任务。

2.4.3 完全分布模式安装

完全分布模式需要将 Spark 目录分文件分发到其他主机并配置 worker 节点，由此快速配置 Spark 集群（需要先安装好 JDK 并配置好从 Master 到 Worker 的 SSH 信任）。具体步骤如下：

步骤01 配置计划表。

集群主机配置如表 2-3 所示。所有主机安装 JDK 在相同目录下。Spark 安装到所有主机相同目录下，如/app/。

表 2-3 集群主机配置表

IP/主机名/虚拟机	软 件 程 序	进　　程
192.168.56.101 server101 CentOS7-101	JDK/Spark SSH 向 server101、server102、server103 免密码登录	Master Worker

(续表)

IP/主机名/虚拟机	软件程序	进程
192.168.56.102 server102 CentOS7-102	JDK/Spark	Worker
192.168.56.103 server103 CentOS7-103	JDK/Spark	Worker

步骤02 准备 3 台 Linux 虚拟机搭建集群环境，完成以下工作：

（这里推荐读者直接使用 2.3.3 节配置好的 Hadoop 完全分布式环境，稍微做些修改，即可快速搭建 Spark 完全分布环境。Spark 完全分布模式使用的 3 台虚拟机环境可以参看图 2-36。）

（1）解压并配置 Spark。

在 server101 上解压 Spark：

```
$ tar -zxvf ~/spark-3.3.1-bin-hadoop3.tgz -C /app/
$ mv spark-3.3.1-bin-hadoop3 spark-3.3.1
```

修改 spark-env.sh 文件，在文件最开始添加 JAVA_HOME 环境变量：

```
$ vim /app/spark-3.3.1/sbin/spark-conf.sh
export JAVA_HOME=/usr/java/jdk1.8.0-361
```

修改 worker 文件，添加所有主机在 worker 节点的名称：

```
$ vim /app/spark-3.3.1/conf/workers
server101
server102
server103
```

使用 scp 将 Spark 目录分发到所有主机相同的目录下：

```
$ scp -r /app/spark-3.3.1  server102:/app/
$ scp -r /app/spark-3.3.1  server103:/app/
```

（2）启动 Spark。在主 Spark 上执行 start-all.sh：

```
$ /app/spark-3.3.1/sbin/start-all.sh
```

启动完成以后，查看 master 主机的 8080 端口，如图 2-44 所示。

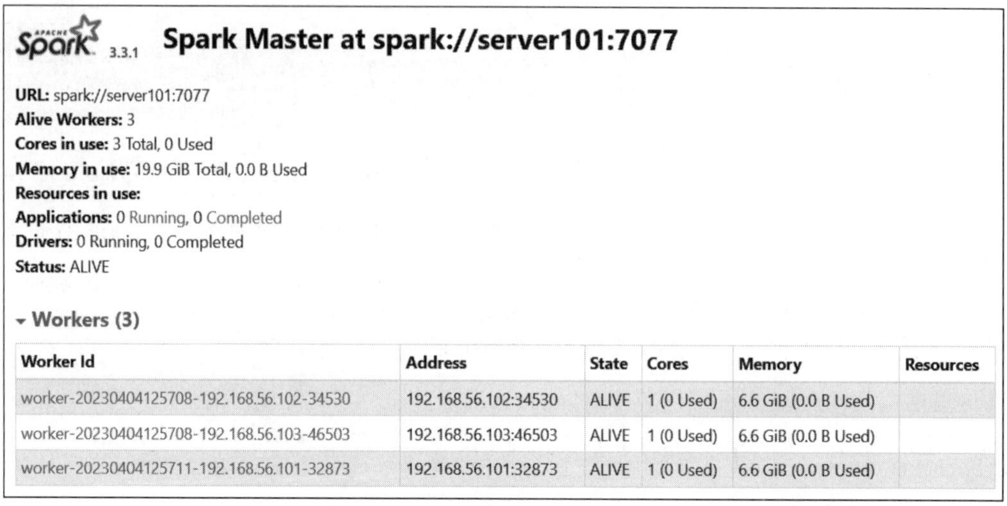

Spark 3.3.1 **Spark Master at spark://server101:7077**

URL: spark://server101:7077
Alive Workers: 3
Cores in use: 3 Total, 0 Used
Memory in use: 19.9 GiB Total, 0.0 B Used
Resources in use:
Applications: 0 Running, 0 Completed
Drivers: 0 Running, 0 Completed
Status: ALIVE

Workers (3)

Worker Id	Address	State	Cores	Memory	Resources
worker-20230404125708-192.168.56.102-34530	192.168.56.102:34530	ALIVE	1 (0 Used)	6.6 GiB (0.0 B Used)	
worker-20230404125708-192.168.56.103-46503	192.168.56.103:46503	ALIVE	1 (0 Used)	6.6 GiB (0.0 B Used)	
worker-20230404125711-192.168.56.101-32873	192.168.56.101:32873	ALIVE	1 (0 Used)	6.6 GiB (0.0 B Used)	

图 2-44

（3）测试。由于已经配置了 Hadoop 集群，且与 Spark 的 worker 节点在相同的主机上，因此在集群环境下，一般是访问 HDFS 上的文件：

```
$spark-shell --master spark://server101:7077
scala> val rdd1 = sc.textFile(“hdfs://server101:8082/test/a.txt”);
```

将结果保存到 HDFS，最后查看 HDFS 上计算的结果即可：

```
scala> rdd1.flatMap(_.split("\\s+")).map((_,1)).reduceByKey(_+_).
saveAsTextFile("hdfs://server101:8020/out004");
```

2.4.4 Spark on YARN

在 Spark Standalone 模式下，集群资源调度由 master 节点负责。Spark 也可以将资源调度交给 YARN 来负责，好处是 YARN 支持动态资源调度。Standalone 模式只支持简单的固定资源分配策略，每个任务固定数量的 core（CPU 核心），各任务按顺序依次分配资源，资源不够时排队等待。这种策略适用于单用户的场景，当有多用户时，各用户的程序差别很大，这种简单粗暴的策略很可能导致有些用户总是分配不到资源，而 YARN 的动态资源分配策略可以很好地解决这个问题。另外，YARN 作为通用的资源调度平台，除了为 Spark 提供调度服务外，还可以为其他子系统（比如 Hadoop MapReduce、Hive）提供调度，这样由 YARN 来统一为集群上的所有计算负载分配资源，可以避免资源分配的混乱无序。

在 Spark Standalone 集群部署完成之后，配置 Spark 支持 YARN 就相对容易多了。

步骤 01 配置 spark-env.sh。

Spark 已经可以配置运行在 YARN 上，只要在 spark-env.sh 中配置 Hadoop 的相关信息和 SPARK_EXECUTOR_CORES 数量即可：

```
$ vim /app/spark-3.3.1/conf/spark-env.sh
HADOOP_CONF_DIR=/app/hadoop-3.2.3/etc/hadoop/
```

```
YARN_CONF_DIR=/app/hadoop-3.2.3/etc/hadoop/
SPARK_EXECUTOR_CORES=2
```

将 Spark 依赖的所有 JAR 包都打包为一个大的 JAR 包，上传到 HDFS 并在 spark-default.sh 中配置这个 JAR 包的位置。进入/app/spark-3.3.1/jars，然后执行打包命令：

```
$ jar -cv0f ~/spark-libs.jar *.jar
```

将打包好的 JAR 包上传到 HDFS：

```
$ hdfs dfs -put ~/spark-libs.jar /spark/spark-libs.jar
```

在 spark-default.sh 中配置上述地址：

```
$ vim /app/spark-3.3.1/conf/spark-default.conf
spark.yarn.archive=hdfs://server101:8020/spark/spark-libs.jar
```

步骤02 启动 spark-shell --master yarn。

使用 spark-shell --master yarn 来启动 Spark 客户端。如果内存不够大，在启动时会出现以下异常：

```
Container is running beyond virtual memory limits. Current usage: 250.2 MB of 1 GB physical memory used;
```

解决方法是取消 YARN 的内存检查，即在 yarn-site.xml 文件中添加以下代码：

```xml
<property>
    <name>yarn.nodemanager.vmem-check-enabled</name>
    <value>false</value>
</property>
```

配置完成以后，重新启动 YARN。

步骤03 测试 Spark on YARN 是否安装成功。

```
[hadoop@server201 app]$ /app/spark-3.3.1/bin/spark-shell --master yarn
2021-03-17 15:49:39 WARN NativeCodeLoader:62 - Unable to load native-hadoop
Spark context Web UI available at http://server201:4040
Spark context available as 'sc' (master = yarn, app id = application_1547711305090_0001).
Spark session available as 'spark'.
Welcome to
      ____              __
     / __/__  ___ _____/ /__
    _\ \/ _ \/ _ `/ __/  '_/
   /___/ .__/\_,_/_/ /_/\_\   version 3.3.1
      /_/

Using Scala version 2.13.8 (Java HotSpot(TM) 64-Bit Server VM, Java 1.8.0_361)
Type in expressions to have them evaluated.
Type :help for more information.
```

Spark on YARN 已安装成功,现在就可以做一个测试,查看 Spark on YARN 的运行结果。

首先在 HDFS 上创建/test/a.txt:

```
[hadoop@server201 ~] hdfs dfs -mkdir /test
[hadoop@server201 ~] hdfs dfs -put ./a.txt /test
```

读取 HDFS 上的一个文件:

```
scala> val rdd1 = sc.textFile("/test/a.txt");
rdd1: org.apache.spark.rdd.RDD[String] = /test/a.txt MapPartitionsRDD[1] at textFile at <console>:24
```

然后统计行数:

```
scala> rdd1.count
res0: Long = 21
```

再执行一系列算子:

```
scala> rdd1.flatMap(_.split("\\s+")).map((_,1)).reduceByKey(_+_).collect.foreach(kv=>println(kv._1+"\t"+kv._2));
Example    1
Program    1
is         1
Hello      1
Word       1
MapReduce  1
This       2
Success!   1
Very       1
Hope       1
Sample     1
Run        1
Count      1
a          1
You        1
of         1
```

执行完成以后,通过浏览器查看 4040 端口和 8088 端口,分别如图 2-45 和图 2-46 所示。当我们访问 4040 端口时,会自动跳转到 8088 端口。

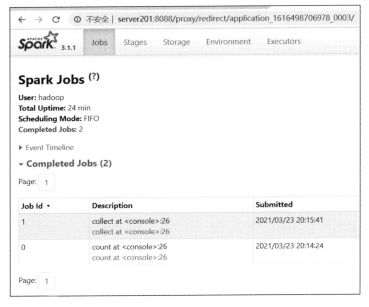

图 2-45

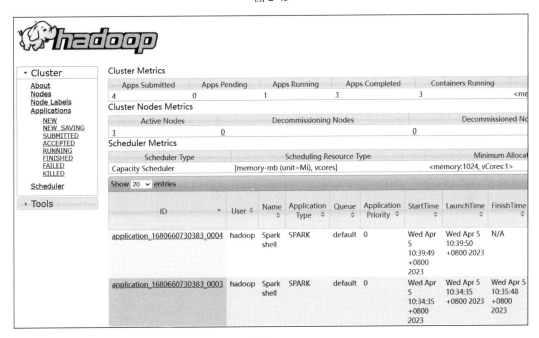

图 2-46

正如本节所介绍的，Spark 的安装与配置非常简单，它的运行也有以下几种方式：

- 本地模式运行，直接使用 spark-shell --master local[*]。
- 伪分布模式和完全分布模式运行（即 Spark Standalone），首先需要使用 start-all.sh 启动 Spark 集群，然后使用 spark-shell --master spark://server:7077 登录 driver。
- 完全分布模式运行在 YARN（Spark on YARN）上，这需要在 spark-env.sh 配置文件中添加 HADOOP_CONF_DIR，然后就可以使用 spark-shell --master yarn 方式来启动 driver。

2.5 spark-submit

本节主要介绍 Spark 官方推荐的提交任务方式——spark-submit。spark-submit 脚本位于 Spark 的 bin 目录中,用于在集群上启动应用程序。这种启动方式可以通过统一的界面使用所有的 Spark 支持的集群管理功能,而用户不必为每个应用程序做专门的配置。

2.5.1 使用 spark-submit 提交

首先查看 spart-submit 的帮助:

```
[hadoop@server201 app]# spark-submit
Usage: spark-submit [options] <app jar | python file | R file> [app arguments]
Usage: spark-submit --kill [submission ID] --master [spark://...]
Usage: spark-submit --status [submission ID] --master [spark://...]
Usage: spark-submit run-example [options] example-class [example args]
```

然后修改一个 Scala 示例程序,输入和输出都通过参数来接收,如代码 2-1 所示。

代码 2-1 WordCount2.scala

```scala
package org.hadoop.spark
object WordCount2 {
  def main(args: Array[String]): Unit = {
    if (args.length < 2) {
      print("usage :<in> <out>");
      return;
    }
    val in: String = args.apply(0);
    val out: String = args.apply(1);
    val conf: SparkConf = new SparkConf();
    conf.set("fs.defaultFS", "hdfs://server201:8020");
    conf.setAppName("WordCount");
    var sc: SparkContext = new SparkContext(conf);
    //获取hadoop config
    val hadoopConfig: Configuration = new Configuration();
    hadoopConfig.set("fs.defaultFS", "hdfs://server201:8020");
    val fs: FileSystem = FileSystem.get(hadoopConfig);
    val pathOut: Path = new Path(out);
    if (fs.exists(pathOut)) {
      fs.delete(pathOut, true);  //删除已经存在的文件
    }
    if (!fs.exists(new Path(in))) {
      print("文件或目录不存在:" + in);
      return;
    }
    val rdd = sc.textFile(in, minPartitions = 2);
    rdd.flatMap(_.split("\\s+"))
      .map((_, 1))
      .reduceByKey(_ + _)
```

```
            .sortByKey()
            .map(kv => kv._1 + "\t" + kv._2)
            .saveAsTextFile(out);
        sc.stop();
    }
}
```

再使用 Maven 打包上述代码，也可以输入命令或是使用 IDEA 打包。

接着将打包好的内容上传到 Linux，再使用以下语句提交代码：

```
$ spark-submit --master spark://server201:7077 \
--class org.hadoop.spark.WordCount2 \
chapter11-1.0.jar\
hdfs://server201:8020/test/   \
hdfs://server201:8020/out001
```

执行完成以后，查看目录下输出的数据即可。

至此我们已经可以使用 Scala 开发 Spark 程序了。

2.5.2　spark-submit 参数说明

Spark 提交任务有两种常见模式：

- local[k]：本地使用 k 个 worker 线程运行 Spark 程序。这种模式适合小批量数据在本地调试代码用（若使用本地的文件，需要在前面加上 file://）。
- Spark on YARN 模式。
    - yarn-client 模式：以 client 模式连接到 YARN 集群，该方式 driver 是在 client 上运行的。
    - yarn-cluster 模式：以 cluster 模式连接到 YARN 集群，该方式 driver 运行在 worker 节点上。

对于应用场景来说，yarn-cluster 模式适合生产环境，yarn-client 模式适合交互和调试。

1. 更多通用可选参数

（1）--master master_url：主机的 URL，参数值可以是 spark://host:port、mesos://host:port、yarn、yarn-cluster、yarn-client、local。

（2）--deploy-mode：driver 程序运行的地方，参数值可以是 client 或者 cluster，默认是 client。

（3）--class：主类名称，含包名。

（4）--jars：逗号分隔的本地 JARS，driver 和 executor 依赖的第三方 JAR 包。

（5）--driver-class-path：驱动依赖的 JAR 包。

（6）--files：用逗号隔开的文件列表，会放置在每个 executor 工作目录中。

（7）--conf：Spark 的配置属性。

例如：

```
spark.executor.userClassPathFirst=true
```

表示当在 executor 中加载类时，用户添加的 JAR 是否比 Spark 自己的 JAR 优先级高。这个属性可以降低 Spark 依赖和用户依赖的冲突。它现在还是一个实验性的特征。

又如：

```
spark.driver.userClassPathFirst=true
```

表示当在 driver 中加载类时，用户添加的 JAR 是否比 Spark 自己的 JAR 优先级高。这个属性可以降低 Spark 依赖和用户依赖的冲突。它现在还是一个实验性的特征。

又如：

```
--driver-memory
```

driver 程序使用内存大小（例如：1000MB，5GB），默认 1024MB。

又如：

```
--executor-memory
```

每个 executor 内存大小（例如：1000MB，2GB），默认 1GB。

2. 仅限于 Spark on YARN 模式的参数

（1）--driver-cores：driver 使用的 core，仅在 cluster 模式下，默认为 1。

（2）--queue：QUEUE_NAME 指定资源队列的名称。默认为 default。

（3）--num-executors：一共启动的 executor 数量，默认是 2 个。

（4）--executor-cores：每个 executor 使用的内核数，默认为 1。

3. 几个重要的参数说明

（1）executor_cores*num_executors：表示的是能够并行执行的任务的数目不宜太小或太大，一般不超过总队列 cores 的 25%。比如队列总 cores 为 400，则最大不要超过 100，最小建议不低于 40，除非日志量很小。

（2）executor_cores：不宜为 1，否则 work 进程中线程数过少，一般 2~4 为宜。

（3）executor_memory：一般 6~10GB 为宜，最大不超过 20GB，否则会导致 GC（垃圾回收）代价过高，或资源浪费严重。

（4）driver-memory：driver 不做任何计算和存储，只是下发任务与 YARN 资源管理器、task 交互，除非是 spark-shell，否则一般为 1~2GB。

4. 增加 executor 的内存量

增加每个 executor 的内存量，增加了内存量以后，对性能的提升有以下 3 点好处：

（1）如果需要对 RDD 进行缓存，那么更多的内存就可以缓存更多的数据，写入更少的数据到磁盘，甚至不写入磁盘，因此减少了磁盘 I/O，提升了性能。

（2）对于 shuffle 操作，在 reduce 端，会需要内存来存放拉取的数据并进行聚合，如果内存不够，那么也会写入磁盘。如果给 executor 分配更多内存，那么就有更少的数据需要写入磁盘，甚至不需要写入磁盘，因此减少了磁盘 I/O，提升了性能。

（3）对于任务的执行，可能会创建很多对象。如果内存比较小，可能会频繁导致 JVM 内存占满,从而导致频繁地进行垃圾回收,影响执行速度。内存加大以后,带来更少的垃圾回收操作,因此避免了执行速度变慢,提升了性能。

Spark 提交参数的设置非常重要,如果设置得不合理,会影响性能,所以要根据具体的情况适当地调整参数的配置,以利于提高程序执行的性能。

2.6 DataFrame

Spark 的运行和计算都围绕 DataFrame 来进行。DataFrame 可以看作一个简单的"数据矩阵(数据框)"或"数据表",对它进行操作也只需要调用有限的数组方法即可。它与一般的"表"的区别在于 DataFrame 是分布式存储,可以更好地利用现有的云数据平台,并在内存中运行。

2.6.1 DataFrame 概述

DataFrame 实质上是存储在不同节点计算机中的一张关系型数据表。分布式存储最大的好处是可以让数据在不同的工作节点(worker)上并行存储,以便在需要数据的时候并行运算,从而获得最迅捷的运行效率。

1. DataFrame 与 RDD 的关系

RDD 是一种分布式弹性数据集,将数据分布在不同节点的计算机内存中进行存储和处理。每次 RDD 对数据处理的最终结果都分别存放在不同的节点中。R 即 Resilient,是弹性的意思,在 Spark 中指的是数据的存储方式,即数据在节点中进行存储时既可以使用内存也可以使用磁盘。这为使用者提供了很大的自由,提供了不同的持久化和运行方法,它是一种具有容错机制的特殊数据集合。

RDD 可以说是 DataFrame 的前身,DataFrame 是 RDD 的发展和拓展。RDD 中可以存储任何单机类型的数据,但是直接使用 RDD 在字段需求明显时存在算子难以复用的缺点。例如,假设 RDD 存放的数据是一个 Person 类型的数据,现在要求出所有年龄段(10 年为一个年龄段)中最高的身高与最大的体重。使用 RDD 接口时,因为 RDD 不了解其中存储的数据的具体结构,所以需要用户自己去写一个很特殊的聚合函数来完成这样的功能。那么如何改进才可以让 RDD 了解其中存储的数据包含哪些列,并在列上进行操作呢?

根据谷歌的解释,DataFrame 是表格或二维数组状结构,其中每一列包含对一个变量的量度,每一行包含一个案例,类似于关系数据库中的表或者 R/Python 中的 dataframe,可以说是一个具有良好优化技术的关系表。

有了 DataFrame,框架会了解 RDD 中的数据具有什么样的结构和类型,使用者可以说清楚自己对每一列进行什么样的操作。这样就有可能实现一个算子,用在多列上比较容易进行算子的复用,甚至在需要同时求出每个年龄段内不同的姓氏有多少个的时候使用 RDD 接口。在之前的函数需要很大的改动才能满足需求时使用 DataFrame 接口,这时只需要添加对这一列的处理,原来的 max/min 相关列的处理都可保持不变。

在 Apache Spark 里,DataFrame 优于 RDD,但也包含了 RDD 的特性。RDD 和 DataFrame 的共同特征是不可变性、内存运行、弹性、分布式计算能力,即 DataFrame=RDD[Row]+shcema。这里尽

量避免理论化的探讨，尽量讲解得深入一些，毕竟本书是以实战为主的。

分布式数据的容错性处理是涉及面较广的问题，较为常用的方法主要有两种：

- 检查节点：对每个数据节点逐个进行检测，随时查询每个节点的运行情况。这样做的好处是便于操作主节点，随时了解任务的真实数据运行情况；坏处是系统进行的是分布式存储和运算，节点检测的资源耗费非常大，而且一旦出现问题，就需要将数据在不同节点中搬运，反而更加耗费时间，从而极大地降低了执行效率。
- 更新记录：运行的主节点并不总是查询每个分节点的运行状态，而是将相同的数据在不同的节点（一般情况下是 3 个）中进行保存，各个工作节点按固定的周期更新在主节点中运行的记录，如果在一定时间内主节点查询到数据的更新状态超时或者存在异常，就在存储相同数据的不同节点上重新启动数据计算工作。其缺点在于数据量过大时，更新数据和重新启动运行任务的资源耗费也相当大。

2. DataFrame 的特性

1）不可变性

DataFrame 是一个不可变的分布式数据集合，与 RDD 不同，DataFrame 就像关系数据库中的表一样，即具有定义好的行、列的分布式数据表。

DataFrame 背后的思想是允许处理大量结构化数据。DataFrame 包含带 schema 的行。schema 是数据结构的说明，意为模式。schema 是 Spark 的 StructType 类型，由一些域（StructFields）组成，域中明确了列名、列类型以及一个布尔类型的参数（表示该列是否可以有缺失值或 null 值），最后还可以可选择地明确该列关联的元数据（在机器学习库中，元数据是一种存储列信息的方式，平常很少用到）。schema 提供了详细的数据结构信息，例如包含哪些列、每列的名称和类型各是什么。DataFrame 由于其表格格式而具有其他元数据，这使得 Spark 可以在最终查询中运行某些优化。

使用一行代码即可输出 schema，代码如下：

```
df.printSchema()
//看看schema到底长什么样子
```

2）延迟计算

DataFrame 的另外一大特性是延迟计算（懒惰执行），即一个完整的 DataFrame 运行任务被分成两部分：Transformation 和 Action（转化操作和行动操作）。转化操作就是从一个 RDD 产生一个新的 RDD，行动操作就是进行实际的计算。只有当执行一个行动操作时，才会执行并返回结果。下面仍然以 RDD 这种数据集为例来解释一下这两种操作。

- Transformation

Transformation 用于创建 RDD。在 Spark 中，RDD 只能使用 Transformation 创建，同时 Transformation 还提供了大量的操作方法，例如 map、filter、groupBy、join 等。除此之外，还可以利用 Transformation 生成新的 RDD，在有限的内存空间中生成尽可能多的数据对象。有一点要牢记，无论发生了多少次 Transformation，在 RDD 中数据计算运行的操作都不可能真正运行。

- Action

Action 是数据的执行部分，通过 count、reduce、collect 等方法真正执行数据的计算部分。实际上，RDD 中所有的操作都是使用 Lazy 模式（一种程序优化的特殊形式）进行的，运行在编译的过程中，不会立刻得到计算的最终结果，而是记住所有的操作步骤和方法，只有显式地遇到启动命令才进行计算。

这样做的好处在于：大部分优化和前期工作在 Transformation 中已经执行完毕，当进行 Action 工作时，只需要利用全部资源完成业务的核心工作。

Spark SQL 可以使用其他 RDD 对象、Parquet 文件、JSON 文件、Hive 表以及通过 JDBC 连接到其他关系数据库作为数据源，来生成 DataFrame 对象。它常用来处理存储系统 HDFS、Hive 表、MySQL 等上面的数据。

2.6.2　DataFrame 的基础应用

1. 创建 DataFrame

Spark 3 推荐使用 SparkSession 来创建 Spark 会话，然后利用使用 SparkSession 创建出来的 Application 来创建 DataFrame。示例如下：

```
import org.apache.spa
val spark = SparkSession
  .builder()                                              //创建 Spark 会话
  .appName("Spark SQL basic example")                     //设置会话名称
  .master("local")                                        //设置本地模式
  .config("spark.some.config.option", "some-value")       //设置相关配置
  .getOrCreate()                                          //创建会话变量
```

对于所有的 Spark 功能，SparkSession 类都是入口，所以创建基础的 SparkSession 只需要使用 SparkSession.builder()。使用 SparkSession 时，应用程序能够从现存的 RDD、Hive 表或者 Spark 数据源里面创建 DataFrame，也可以直接将数据源里读成 DataFrame 的格式。

以下是完整的 createDataFrame 方法：

```
import org.apache.spark.sql._
import org.apache.spark.sql.types._
val sparkSession = new org.apache.spark.sql.SparkSession(sc)
val schema =
  StructType(
    StructField("name", StringType, false) ::
    StructField("age", IntegerType, true) :: Nil)
val people =
  sc.textFile("examples/src/main/resources/people.txt").map(
    _.split(",")).map(p => Row(p(0), p(1).trim.toInt))
val dataFrame = sparkSession.createDataFrame(people, schema)
dataFrame.printSchema
//root
//|-- name: string (nullable = false)
//|-- age: integer (nullable = true)
```

```
dataFrame.createOrReplaceTempView("people")
sparkSession.sql("select name from people").collect.foreach(println)
```

代码解释：

从上面代码中可以看到，createDataFrame 方法在使用时是借助 SparkSession 会话环境进行工作的，因此需要对 Spark 会话环境变量进行设置。以上代码先从一个文件里创建一个 RDD，再使用 createDataFrame 方法，其中第一个参数是 RDD、第二个参数 schema 是上面定义的 DataFrame 的字段数据类型等信息。

另一种创建 DataFrame 的方式是使用 toDF() 函数将强类型数据集合转换为带有重命名列的通用 DataFrame。这在将元组的 RDD 转换为具有有意义名称的 DataFrame 时非常方便，示例如下：

```
import spark.implicits._
val rdd: RDD[(Int, String)] = ...
rdd.toDF()    //这里是一个隐式转换，将 RDD 变成列名为_1、_2 的 DataFrame
rdd.toDF("id", "name")    //将 RDD 变成列名为"id"、"name"的 DataFrame
//需要添加结构信息并加上列名
```

提示： 对 DataFrame、DataSet 和 RDD 进行转换需要在代码中使用 import spark.implicits._ 以获得这个包的支持。

2. select 和 selectExpr 方法

select 和 selectExpr 方法用于把 DataFrame 中的某些列筛选出来。其中，select 用来选择某些列出现在结果集中，并且结果作为一个新的 DataFrame 返回，使用方法如下：

```
import org.apache.spark.sql.SparkSession
object select {
    def main(args: Array[String]): Unit = {
        val spark = SparkSession
            .builder()                                    //创建 Spark 会话
            .appName("Spark SQL basic example")           //设置会话名称
            .master("local")                              //设置本地模式
            .getOrCreate()                                //创建会话变量
        val rdd = spark.sparkContext.parallelize(Array(1,2,3,4))
        import spark.implicits._
        val df = rdd.toDF("id")
        df.select("id").show()                            //选择"id"列
    }
}
```

打印结果如下：

```
+---+
| id|
+---+
|  1|
|  2|
|  3|
|  4|
```

如果是 selectExpr 方法,那么代码如下:

```
import org.apache.spark.sql.SparkSession
object select {
  def main(args: Array[String]): Unit = {
    val spark = SparkSession
      .builder()                                      //创建 Spark 会话
      .appName("Spark SQL basic example")             //设置会话名称
      .master("local")                                //设置本地模式
      .getOrCreate()                                  //创建会话变量
    val rdd = spark.sparkContext.parallelize(Array(1,2,3,4))
    import spark.implicits._
    val df = rdd.toDF("id")
    df.selectExpr("id as ID").show()                  //设置了一个别名 ID
  }
}
```

具体结果请读者自行运行查看。

3. collect 方法

collect 方法将已经存储的 DataFrame 数据从存储器中收集回来,并返回一个数组,包括 DataFrame 集合中所有的行,其格式如下:

```
def collect(): Array[T]
```

Spark 的数据是分布式存储在集群上的,如果想获取一些数据在本机 Local 模式上进行操作,就需要将数据收集到 driver 驱动器中。collect() 返回 DataFrame 中的全部数据,并返回一个 Array 对象,示例代码如下:

```
import org.apache.spark.sql.SparkSession
object collect {
  def main(args: Array[String]): Unit = {
    val spark = SparkSession
      .builder()                                      //创建 Spark 会话
      .appName("Spark SQL basic example")             //设置会话名称
      .master("local")                                //设置本地模式
      .getOrCreate()                                  //创建会话变量
    val rdd = spark.sparkContext.parallelize(Array(1,2,3,4))
    import spark.implicits._
    val df = rdd.toDF("id")
    val arr = df.collect()
    println(arr.mkString("Array(", ", ", ")"))
  }
}
```

注意:将数据收集到驱动器中,尤其是当数据集很大或者分区数据集很大时,很容易让驱动器崩溃;数据收集到驱动器中进行计算,就不是分布式并行计算了,而是串行计算,速度会更慢,所以除了查看小数据,一般不建议将数据收集到驱动器中。

4. 计算行数 count 方法

count 方法用来计算数据集 DataFrame 中行的个数，返回 DataFrame 集合的行数，使用方法如下：

```
import org.apache.spark.sql.SparkSession
object count {
   def main(args: Array[String]): Unit = {
      val spark = SparkSession
        .builder()                                    //创建 Spark 会话
        .appName("Spark SQL basic example")           //设置会话名称
        .master("local")                              //设置本地模式
        .getOrCreate()                                //创建会话变量
      val rdd = spark.sparkContext.parallelize(Array(1,2,3,4))
      import spark.implicits._
      val df = rdd.toDF("id")
      println(df.count())                             //计算行数
   }
}
```

最终结果如下：

4

5. filter 方法

filter 方法是一个比较常用的方法，用来按照条件过滤数据集。如果想选择 DataFrame 的某列中大于或小于某某的数据，就可以使用 filter 方法。对于多个条件，可以将 filter 方法写在一起。

filter 方法接收任意一个函数作为过滤条件。行过滤的逻辑是先创建一个判断条件表达式，根据表达式生成 true 或 false，然后过滤使表达式值为 false 的行。filter 方法的使用示例代码如下：

```
import org.apache.spark.sql.SparkSession
object fliter {
   def main(args: Array[String]): Unit = {
      val spark = SparkSession
        .builder()                                    //创建 Spark 会话
        .appName("Spark SQL basic example")           //设置会话名称
        .master("local")                              //设置本地模式
      .getOrCreate()                                  //创建会话变量
      val rdd = spark.sparkContext.parallelize(Array(1,2,3,4))
      import spark.implicits._
      val df = rdd.toDF("id")
      val df2 = df.filter("id>3")//过滤 id 列中大于 3 的数据（行）
      println(df2.cache().show())                     //打印结果
   }
}
```

具体结果请读者自行验证。这里需要说明的是，"_>=3"采用的是 Scala 编程中的规范，"_"的作用是作为占位符标记所有传过来的数据，在此方法中，数组的数据（1、2、3、4）依次传递进来替代了占位符。

6. flatMap 方法

flatMap 方法是对 DataFrame 中的数据集进行整体操作的一个特殊方法，因为它在定义时是针对数据集进行操作的，因此最终返回的也是一个数据集。flatMap 方法首先将函数应用于此数据集中的所有元素，然后将结果展平，从而返回一个新的数据集。示例代码如下：

```
import org.apache.spark.sql.SparkSession
object flatmap {
   def main(args: Array[String]): Unit = {
      val spark = SparkSession
        .builder()                                      //创建Spark会话
        .appName("Spark SQL basic example")             //设置会话名称
        .master("local")                                //设置本地模式
        .getOrCreate()                                  //创建会话变量
      val rdd = spark.sparkContext.parallelize(Seq("hello!spark",
"hello!hadoop"))
      import spark.implicits._
      val df = rdd.toDF("id")
      val x = df.flatMap(x => x.toString().split("!")).collect()
      println(x.mkString("Array(", ", ", ")"))
   }
}
```

7. groupBy 和 agg 方法

groupBy 方法是将传入的数据进行分组，分组依据是作为参数传入的计算方法。聚合操作调用的是 agg 方法，该方法有多种调用方式，一般与 groupBy 方法配合使用。在使用 groupBy 时，通常都是先分组再使用 agg 等聚合函数对数据进行聚合。groupBy+agg 的使用示例如下：

```
import org.apache.spark.sql.SparkSession
object groupBy {
   def main(args: Array[String]): Unit = {
      val spark = SparkSession
        .builder()                                      //创建Spark会话
        .appName("Spark SQL basic example")             //设置会话名称
        .master("local")                                //设置本地模式
        .getOrCreate()                                  //创建会话变量
      val df = spark.read.json("./src/C03/people.json")
      df.groupBy("name").agg("age" -> "count").show()
   }
}
```

这里采用 groupBy+agg 的方法统计了 age 字段的条数。

在 GroupedData 的 API 中提供了 groupBy 之后的操作，比如：

- max(colNames: String*)方法：获取分组中指定字段或者所有的数字类型字段的最大值，只能作用于数字型字段。
- min(colNames: String*)方法：获取分组中指定字段或者所有的数字类型字段的最小值，只能作用于数字型字段。

- mean(colNames: String*)方法：获取分组中指定字段或者所有的数字类型字段的平均值，只能作用于数字型字段。
- sum(colNames: String*)方法：获取分组中指定字段或者所有的数字类型字段的和值，只能作用于数字型字段。
- count 方法：获取分组中的元素个数。

以上这些方法都等同于 agg 方法。

8. drop 方法

drop 方法从数据集中删除某列，然后返回 DataFrame 类型，示例代码如下：

```
import org.apache.spark.sql.SparkSession
object drop {
  def main(args: Array[String]): Unit = {
    val spark = SparkSession
      .builder()                                      //创建 Spark 会话
      .appName("Spark SQL basic example")             //设置会话名称
      .master("local")                                //设置本地模式
      .getOrCreate()                                  //创建会话变量
    val df = spark.read.json("./src/C03/people.json")
    df.drop("age").show()                             //删除 age 列
  }
}
```

最终打印结果如下：

```
+-------+
|   name|
+-------+
|Michael|
|   Andy|
| Justin|
+-------+
```

这里也可以通过 select 方法来实现列的删除，不过建议使用专门的 drop 方法来实现——规范又显而易见，对于维护工作来说最有效率。

2.7 Spark SQL

Spark SQL 是 Spark 的一个模块，主入口是 SparkSession，将 SQL 查询与 Spark 程序进行无缝混合。DataFrame 和 SQL 提供了访问各种数据源（通过 JDBC 或 ODBC 连接）的常用方法，包括 Hive、Avro、Parquet、ORC、JSON 和 JDBC。我们可以使用相同方式连接到任何不同的数据源。Spark SQL 还支持 HiveQL 语法以及 Hive SerDes、UDF，允许我们访问现有的 Hive 仓库。

Spark SQL 包括基于成本的优化器、列式存储和代码生成，以快速进行查询。同时，它使用 Spark 引擎扩展到数千个节点和多小时查询，该引擎提供完整的中间查询容错。不要担心使用不同的引擎来获取历史数据时出现问题。

2.7.1 快速示例

1. 将 RDD 转换成 DataFrame

首先创建一个 RDD：

```
scala> val rdd=sc.makeRDD(Seq(("Jack",24),("Mary",34)));
```

再转成 DataFrame：

```
scala> val df1 = rdd.toDF();
df1: org.apache.spark.sql.DataFrame = [_1: string, _2: int]
```

然后使用 show 显示数据：

```
scala> df1.show();
+----+---+
|  _1| _2|
+----+---+
|Jack| 24|
|Mary| 34|
+----+---+
```

2. 给 DataFrame 设置别名

代码如下：

```
scala> val df2 = rdd.toDF("name","age");
```

再次使用 show 显示时，将显示列的名称：

```
scala> df2.show();
+----+---+
|name|age|
+----+---+
|Jack| 24|
|Mary| 34|
+----+---+
```

3. 使用 SqlContext 执行 SQL 语句

首先保存成 view：

```
scala> df2.createOrReplaceTempView("person");
```

再声明 SqlContext 对象：

```
scala> val sqlContext = spark.sqlContext
```

然后执行 SQL 语句：

```
scala> sqlContext.sql("select * from person").show();
+----+---+
|name|age|
```

```
+----+---+
|Jack| 24|
|Mary| 34|
+----+---+
```

4. Scala 代码将 RDD 转成 Bean

首先在 Scala 项目中将 RDD 转成 Bean，需要添加 spark-sql_2.12 的依赖。

```
<dependency>
    <groupId>org.apache.spark</groupId>
    <artifactId>spark-sql_2.12</artifactId>
    <version>3.3.1</version>
</dependency>
```

然后准备一个文本文件，一行为一个对象，每一个值之间用空格分开，文件名为"D:/a/stud.txt"，内容如下：

```
4 Jack 34 男
1 Mike 23 女
2 刘长友 45 男
3 雪丽 27 女
```

接下来开发以下代码，读取 stud.txt 文件的内容并封闭到 class Stud 对象中。

代码 2-2　RddToBean.scals

```scala
package org.hadoop.spark
object RddToBean {
    def main(args: Array[String]): Unit = {
        val conf: SparkConf = new SparkConf();
        conf.setMaster("local[*]");
        conf.setAppName("RDDToBean");
        val spark: SparkSession =
SparkSession.builder().config(conf).getOrCreate();
        val sc: SparkContext = spark.sparkContext;
        //读取文件
        import spark.implicits._;
        //步 2：读取文件
        val rdd: RDD[String] = sc.textFile("file:///D:/a/stud.txt");
        //步 3：第一次使用 map 对每一行进行切分，第二次使用 map 将数据封装到 Bean 中，最后使用 toDF 转换成 DataFrame
        val df = rdd.map(_.split("\\s+")).map(arr => {
            Stud(arr(0).toInt, arr(1), arr(2).toInt, arr(3));
        }).toDF();
        //步 4：显示或是保存数据
        df.show();
        spark.close();
    }
    /** 步 1：声明 JavaBean，并直接声明主构造方法 * */
    case class Stud(id: Int, name: String, age: Int, sex: String) {
        /** 声明无参数的构造，调用主构造方法 * */
```

```
        def this() = this(-1, null, -1, null);
    }
}
```

最后直接在 IDEA 中运行,输出结果如下:

```
+---+------+---+---+
| id|  name|age|sex|
+---+------+---+---+
|  4|  Jack| 34| 男|
|  1|  Mike| 23| 女|
|  2|刘长友| 45| 男|
|  3|  雪丽| 27| 女|
+---+------+---+---+
```

5. WordCount 示例

让我们再来做一次 WordCount 示例,不过本次使用的是 Spark SQL。

首先读取文件 1.txt,1.txt 中可以是任意的内容:

```
scala> val rdd = sc.textFile("file:///home/hadoop/1.txt");
```

然后以空白字符进行分割并转换成 DF 对象,注意转换后的对象只有一个字段 str。

```
scala> val df3 = rdd.flatMap(_.split("\\s+")).toDF("str");
```

现在可以直接使用 groupBy 进行 count 计算:

```
scala> df3.groupBy("str").count().show();
+------+-----+
|   str|count|
+------+-----+
|     A|    3|
|     B|    1|
|     C|    1|
...
```

还可以指定排序规则:

```
scala> df3.groupBy("str").count().sort("str").show();
+-----+-----+
|  str|count|
+-----+-----+
|   ->|    4|
|    0|    2|
...
```

或是直接使用 SQL 语句:

```
scala> df3.createOrReplaceTempView("words");
scala> sqlContext.sql("select count(str),str from words group by str").show();
+----------+-----+
|count(str)|  str|
+----------+-----+
```

```
|    3|   7|
|    1| lib|
...
```

6. Scala 代码示例

开发 Scala 代码对 stud.txt 文件内容执行统计计算，代码如下：

代码 2-3　SparkSQL.scala

```
package org.hadoop.spark
object SparkSQL {
    def main(args: Array[String]): Unit = {
        val conf: SparkConf = new SparkConf();
        conf.setMaster("local[2]");
        conf.setAppName("SQL");
        val session: SparkSession = 
SparkSession.builder().config(conf).getOrCreate();
        val sqlContext: SQLContext = session.sqlContext;
        val ctx: SparkContext = session.sparkContext;
        val rdd: RDD[String] = ctx.textFile("file:///D:/a/stud.txt");
        //注意要做隐式导入
        import session.implicits._;
        val df: DataFrame = rdd.flatMap(_.split("\\s+")).toDF("str");
        df.show();
        df.groupBy("str").count().sort("str").show();
        //创建View
        df.createTempView("words");
        sqlContext.sql("select str,count(str) cnt from words group by str order by str") //执行 SQL
            //转成 RDD
            .rdd
            .saveAsTextFile("file:///D:/a/out001");  //保存到指定目录
        session.close();
    }
}
```

本地测试成功后再打包到集群上运行，修改后的代码如下：

代码 2-4　SparkSQL2.scala

```
package org.hadoop.spark
object SparkSQL2 {
    def main(args: Array[String]): Unit = {
        if (args.length != 2) {
            println("usage : in out");
            return;
        }
        val inPath: String = args.apply(0);
        val outPath: String = args.apply(1);
        val hConf: Configuration = new Configuration();
        val fs: FileSystem = FileSystem.get(hConf);
```

```
        val dest: Path = new Path(outPath);
        if (fs.exists(dest)) {
            fs.delete(dest, true);
        }
        val conf: SparkConf = new SparkConf();
        conf.setAppName("SQL");
        val session: SparkSession =
SparkSession.builder().config(conf).getOrCreate();
        val sqlContext: SQLContext = session.sqlContext;
        val ctx: SparkContext = session.sparkContext;
        val rdd: RDD[String] = ctx.textFile(inPath);
        //注意要做隐式导入
        import session.implicits._;
        val df: DataFrame = rdd.flatMap(_.split("\\s+")).toDF("str");
        df.show();
        df.groupBy("str").count().sort("str").show();
        df.createTempView("words"); //创建 View
        sqlContext.sql("select str,count(str) cnt from words group by str order by str") //执行 SQL
            .rdd //转成 RDD
            .saveAsTextFile(outPath); //保存到指定目录
        session.close();
    }
}
```

使用 spark-submit 提交任务:

```
# spark-submit --master spark://server201:7077 --class
org.hadoop.spark.SparkSQL2 chapter1-1.0.jar hdfs://server201:8020/test/
hdfs://server201:8020/out002
```

2.7.2　read 和 write

DataFrame/DataSet 提供了 read 和 write 方法,其中 read 可以读取指定格式的数据,而 write 可以写出指定格式的数据。

1. read 方法

read 方法可以读取的数据格式包含:csv、format、jdbc、json、load、option、options、orc、parquet、schema、table、text、textFile。

在没有设置 spark.sql.sources.default 的情况下,默认读取的是 Parquet 格式的文件,由于 Hive 在创建表时可以通过 stored as parquet 存储数据为 Parquet 格式,所以 Spark 可以读取 Hive 的表数据进行查看。

首先,在 Hive 上创建一个 Parquet 存储格式的表,使用 insert 写入一些记录:

```
Hive> create table teacher (id integer,name string ) stored as parquet;
Hive> insert into teacher values(1,'Mary');
Hive> insert into teacher values(2,'刘长秒');
Hive> insert into teacher values(3,'Alex');
```

接下来就可以直接使用 Spark 读取 Parquet 格式的数据了。

（1）使用 read 来读取：

```
scala> spark.read.load("hdfs://server201:8020/user/hive/warehouse/one.db/teacher/*");
res5: org.apache.spark.sql.DataFrame = [id: int, name: string]
```

然后就可以直接显示了：

```
scala> res5.show();
+---+------+
| id|  name|
+---+------+
|  2|刘长秒|
|  1|  Mary|
|  3|  Alex|
+---+------+
```

（2）textFile 用于读取文本文件，它会将整个文件所有行作为一个字段来处理：

```
scala> spark.read.textFile("hdfs://server201:8020/user/hive/warehouse/one.db/stud/stud.txt");
res7: org.apache.spark.sql.Dataset[String] = [value: string]
scala> res7.show();
+---------+
|    value|
+---------+
|   1 Jack|
...
```

2. write 方法

DataFrame/Dataset 的 write.save 方法默认将数据保存成 Parquet 形式。

下面示例将一个 DataFrame 的数据保存到 HDFS 上。

首先是声明：

```
scala> val rdd = sc.parallelize(Seq((1,"Jack"),(2,"Mary")));
```

然后转换成 DataFrame：

```
scala> val df = rdd.toDF("id","name");
```

最后是保存：

```
scala> df.write.save("hdfs://server201:8020/out001");
```

以下保存成 JSON 格式：

```
scala> df2.write.format("json").save("hdfs://server21:8020/out003");
```

2.8 Spark Streaming

Spark Streaming 是 Spark API 核心的扩展，支持可扩展、高吞吐量、实时数据流的容错流处理。数据可以从 Kafka、Flume、Kinesis 或 TCP Socket 等许多来源摄入，并且可以使用复杂的算法和高级别功能 map、reduce、join 和 window 来处理。最后，处理的数据可以推送到文件系统、数据库和实时（dashborad）仪表板。事实上，我们可以将 Spark 的机器学习（Machine Learning，ML）和图形处理（Graphx）算法应用于数据流。如图 2-47 所示为 Spark Streaming 的官方图例。

图 2-47

Spark Streaming 提供了一个高层次的抽象，被称为离散流（DStream），它代表连续的数据流。DStream 可以通过 Kafka、Flume 和 Kinesis 等来源的输入数据流创建，也可以通过在其他 DStream 上应用高级操作来创建。在内部表示为一系列的 RDD。

接下来我们根据官网的示例来快速演示一个 Spark Streaming 程序。此程序将监听某个端口的数据，并将接收到的数据输出到控制台。

步骤01 开发 Spark Streaming 程序，首先需要添加 Spark Streaming 的依赖：

```xml
<dependency>
    <groupId>org.apache.spark</groupId>
    <artifactId>spark-streaming_2.12</artifactId>
    <version>3.3.1</version>
</dependency>
```

步骤02 开发 Scala 程序，代码如下：

代码 2-5　Streaming.java

```scala
package org.hadoop.spark
object Streaming {
    def main(args: Array[String]): Unit = {
        val conf = new SparkConf()
          .setMaster("local[2]") //至少两个线程
          .setAppName("NetworkWordCount")
        //声明 SparkStreaming 对象
        val ssc = new StreamingContext(conf, Seconds(2))
        //声明监听的服务器及端口
        val lines = ssc.socketTextStream("192.168.56.201", 9999)
        //输出接收到的这个端口的数据
        lines.print();
        //开始运行
        ssc.start();
        ssc.awaitTermination();
```

     }
}
```

**步骤03** 向 9999 端口发送数据。

在 Linux 上安装 Netcat 软件：

```
$ sudo yum install -y nc
```

启动 9999 端口：

```
$ nc -lk 9999
```

向此端口输入一些数据：

```
Jack Mary
```

步骤04 启动 Scala 程序并查看输出。

程序启动以后，每 2 秒会输出一次访问数据，如果访问端口时获取到这些数据，就输出数据到控制台，如下所示：

```
-------------------------------------------
Time: 1547961188000 ms
-------------------------------------------
Jack Mary
```

也可以将 lines.print() 语句修改成以下代码，即对数据进行处理计算 WordCount：

```
lines.flatMap(_.split("\\s+")).map((_, 1)).reduceByKey(_ + _).print();
```

步骤05 打包运行。

打包方式参照 1.1 节。

打包以后，通过 spark-submit 提交运行，可以得到相同的效果。

```
# spark-submit --master spark://server201:7077 \
--class org.hadoop.spark.Streaming chapter1-1.0.jar
```

步骤06 打包运行在 Spark 集群上。

对于使用 spark-submit 提交的代码，如果 Spark 集群中仅有一个 worker，那么 Spark Streaming 无法运行，所以，必须保证 worker 节点大于或等于 2 个。

修改代码 2-5，删除 conf.setMaster("local[2]")后，默认为 conf.setMaster("spark://server101:7077");，完整代码如下：

代码 2-6 Streaming2.scala

```
package org.hadoop.spark
import org.apache.spark.SparkConf
import org.apache.spark.streaming.{Seconds, StreamingContext}
object Streaming2 {
    def main(args: Array[String]): Unit = {
        val conf: SparkConf = new SparkConf();
        conf.set("fs.defaultFS", "hdfs://server201:8020");
```
```

```
        conf.setAppName("Streaming01");
        val streamingContext: StreamingContext = new StreamingContext(conf,
Seconds(5));
        var lines = streamingContext.socketTextStream("server101", 4444);
        lines.print();
        lines.saveAsTextFiles("hdfs://server201:8020/out001/a");
        streamingContext.start();
        streamingContext.awaitTermination();
    }
}
```

启动 Spark 集群，查看 worker 个数，如图 2-48 所示。

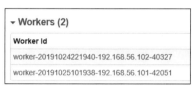

图 2-48

执行 nc 命令：

```
$ nc -lk 4444
```

打包并启动 Spark：

```
$ spark-submit --master spark://server201:7077 \
--class org.hadoop.spark.Streaming2  chapter1-1.0.jar
```

启动后，查看 cores 的个数，结果为 2，因为有 2 个 worker，所以 cores 为 2，如图 2-49 所示。

图 2-49

输入并接收数据，查看输出结果：

```
-------------------------------------------
Time: 1571972220000 ms
-------------------------------------------
Rose
Hello
```

查看 HDFS 上的数据，如图 2-50 所示。

图 2-50

2.9 共享变量

前面学习了 Spark 的主要抽象 RDD（用于并行操作的两种抽象之一），本节讲解一下 Spark 用于并行操作的第二种抽象——共享变量。默认情况下，当 Spark 在不同节点上作为一组任务并行运行一个函数时，它会将函数中使用的每个变量的副本发送给每个任务。有时，一个变量需要在任务之间共享，或者在任务和驱动程序之间共享。共享变量是指在各个 executor 上都可以访问的变量。共享变量可以让所有 executor 更高效地访问程序共同使用的变量。Spark 支持两种类型的共享变量：广播变量和累加器。

2.9.1 广播变量

广播变量（Broadcast Variable）被序列化以后，传递给每一个 executor，并缓存在 executor 里面，以后在需要访问时，高效地获取广播变量的数据。

通过 SparkContext 的 broadcast() 方法创建广播变量，它返回 BroadCast[T] 类型。

需要注意的是，对于广播变量 BroadCast[T] 类型，如果希望访问它的数据，那么需要调用它的 value 属性。

同时，广播变量是单向传递的，即它只能从 driver 到 executor，所以一个广播变量是没有办法更新的。如果希望更新，就使用累加器（accumulator）。

以下是没有使用广播变量的代码：

代码 2-7　ShareVar.scala

```scala
package org.hadoop.spark
import org.apache.spark.{SparkConf, SparkContext}
object ShareVar {
  def main(args: Array[String]): Unit = {
    val conf:SparkConf = new SparkConf();
    conf.setMaster("local[*]");
    conf.setAppName("ShareVariable");
    val sc:SparkContext = new SparkContext(conf);
    sc.setLogLevel("WARN");
    //声明变量, Map(key)其中 key 参数, 是指根据 key 值查询 value 的数据
    val lookup = Map(1->"a",2->"b",3->"c",4->"d");
    val rdd = sc.parallelize(Seq(1,2,4));
    val rdd2 = rdd.map(lookup(_));
    println(rdd2.collect().toSet);
    sc.stop();
  }
}
```

现在将 lookup 变量声明为广播变量，让程序更高效地运行，代码如下：

代码 2-8　ShareVar2.scala

```scala
package org.hadoop.spark
import org.apache.spark.broadcast.Broadcast
import org.apache.spark.{SparkConf, SparkContext}
object ShareVar2 {
```

```scala
def main(args: Array[String]): Unit = {
  val conf:SparkConf = new SparkConf();
  conf.setMaster("local[*]");
  conf.setAppName("ShareVariable");
  val sc:SparkContext = new SparkContext(conf);
  sc.setLogLevel("WARN");
  //声明广播变量
  val lookup:Broadcast[Map[Int,String]] =
    sc.broadcast[Map[Int,String]](Map(1->"a",2->"b",3->"c",4->"d"));
  val rdd = sc.parallelize(Seq(1,2,4));
  //访问广播变量的值,要使用.value
  val rdd2 = rdd.map(lookup.value(_));
  println(rdd2.collect().toSet);
  sc.stop();
}
}
```

2.9.2 累加器

累加器是任务中只能对它做加法操作的共享变量,类似于 MapReduce 中的计数器。当作业完成以后,driver 可以获取累加器的最终值。

通过 SparkContext 的 accumulator 来初始化一个累加器,如代码 2-9 所示。

代码 2-9 Accumulator.scala

```scala
package org.hadoop.spark
object Accumulator {
  def main(args: Array[String]): Unit = {
    val conf: SparkConf = new SparkConf();
    conf.setMaster("local[2]");
    conf.setAppName("Accumulator");
    val sc: SparkContext = new SparkContext(conf);
    sc.setLogLevel("WARN");
    //声明累加器
    val acc: LongAccumulator = sc.longAccumulator("A");
    val rdd = sc.parallelize(1 to 10);
    val rdd2 = rdd.map(i => {//注意 map 是惰性操作
      acc.add(1);//对累加器加 1
      i + 1;
    });
    println(rdd2.collect().toSet);
    println("输出结果: " + acc.count);  //获取值=10
    sc.stop();
  }
}
```

第 3 章

Spark RDD 弹性分布式数据集

Hadoop 中的 MapReduce 虽然具有自动容错、平衡负载和可拓展性的优点，但是它最大缺点是采用非循环式的数据流模型，使得其在迭代计算时需要进行大量的磁盘 I/O 操作。Spark 中的 RDD 可以很好地解决了这一缺点。我们可以将 RDD 理解为一个分布式存储在集群中的大型数据集合，不同 RDD 之间可以通过转换操作形成依赖关系并实现管道化，从而避免了中间结果的 I/O 操作，提高数据处理的速度和性能。本章将对 RDD 进行详细讲解。

本章主要知识点：

- RDD 概述
- RDD 主要属性
- RDD 的特点
- RDD 的创建与处理过程
- 常见的转换算子和行动算子

3.1 什么是 RDD

RDD（Resilient Distributed Dataset，弹性分布式数据集）是一个不可变的分布式对象集合，是 Spark 中最基本的数据抽象。在代码中 RDD 是一个抽象类，代表一个弹性的、不可变、可分区、里面的元素可并行计算的集合。

每个 RDD 都被分为多个分区，这些分区运行在集群中的不同节点上。RDD 可以包含 Python、Java、Scala 中任意类型的对象，甚至可以包含用户自定义的对象。RDD 的转化操作都是惰性求值的，所以我们不应该把 RDD 看作存放着特定数据的数据集，而最好把每个 RDD 当作我们通过转化操作构建出来的、记录如何计算数据的指令列表。

RDD 表示只读的分区的数据集，对 RDD 进行改动，只能通过 RDD 的转换操作，由一个 RDD

得到一个新的 RDD，新的 RDD 包含了从其他 RDD 衍生所必需的信息。RDD 之间存在依赖，RDD 的执行是按照依赖关系延时计算的。如果依赖关系较长，那么可以通过持久化 RDD 来切断依赖关系。RDD 逻辑上是分区的，每个分区的数据抽象存在，计算的时候会通过一个 compute 函数得到每个分区的数据。如果 RDD 是通过已有的文件系统构建的，那么 compute 函数读取指定文件系统中的数据；如果 RDD 是通过其他 RDD 转换而来的，那么 compute 函数将首先执行转换逻辑，也就是将其他 RDD 的数据进行转换。

3.2 RDD 的主要属性

RDD 的主要属性如下：

- A list of partitions：多个分区。

分区可以看作数据集的基本组成单位。对于 RDD 来说，每个分区都会被一个计算任务处理，并决定了并行计算的粒度。用户可以在创建 RDD 时指定 RDD 的分区数，如果没有指定，就会采用默认值。默认值就是程序所分配到的 CPU Core 的数目。每个分配的存储是由 BlockManager 实现的。每个分区都会被逻辑映射成 BlockManager 的一个 Block，而这个 Block 会被一个 task 负责计算。

- A function for computing each split：计算每个切片（分区）的函数。

Spark 中 RDD 的计算是以分区为单位的，每个 RDD 都会实现 compute 函数以达到这个目的。

- A list of dependencies on other RDDs：与其他 RDD 之间的依赖关系。

RDD 的每次转换都会生成一个新的 RDD，所以 RDD 之间会形成类似于流水线一样的前后依赖关系。在部分分区数据丢失时，Spark 可以通过这个依赖关系重新计算丢失的分区数据，而不是对 RDD 的所有分区进行重新计算。

- Optionally，a Partitioner for key-value RDDs (e.g. to say that the RDD is hash-partitioned)：对存储键-值对的 RDD 来说，还有一个可选的分区器。

只有存储键-值对的 RDD，才会有分区器；没有存储键-值对的 RDD，其分区器的值是 None。分区器不但决定了 RDD 的本区数量，也决定了父 RDDShuffle 输出时的分区数量。

- Optionally，a list of preferred locations to compute each split on (e.g. block locations for an HDFS file)：存储每个切片优先位置的列表。

比如对于一个 HDFS 文件来说，这个列表保存的就是每个分区所在文件块的位置。按照"移动数据不如移动计算"的理念，Spark 在进行任务调度的时候，会尽可能地将计算任务分配到它所要处理的数据块的存储位置。

3.3　RDD 的特点

一个 RDD 可以简单地理解为一个分布式的元素集合。在 Spark 中，所有的工作要么是创建 RDD，要么是转换已经存在的 RDD 成为新的 RDD，要么在 RDD 上执行一些操作来得到一些计算结果。本节主要对 RDD 的一些特点进行讲解。

3.3.1　弹性

RDD 的弹性特点主要表现在以下 4 个方面：

- 存储的弹性：内存与磁盘的自动切换。
- 容错的弹性：数据丢失可以自动恢复。
- 计算的弹性：计算出错重试机制。
- 分片的弹性：可根据需要重新分片。

3.3.2　分区

在分布式程序中，网络通信的开销很大，因此控制数据分布以获得最少的网络传输可以极大地提升程序的整体性能。Spark 程序可以通过控制 RDD 分区的方式来减少通信开销。Spark 中所有的 RDD 都可以进行分区，系统会根据一个针对键的函数对元素进行分区。虽然 Spark 不能控制每个键具体划分到哪个节点上，但是可以确保相同的键出现在同一个分区上。

RDD 的分区原则是分区的个数尽量等于集群中的 CPU 核心（Core）数目。对于不同的 Spark 部署模式而言，都可以通过设置 spark.default.parallelism 这个参数值来配置默认的分区数目。

Spark 框架为 RDD 提供了两种分区方式，分别是哈希分区（HashPartitioner）和范围分区（RangePartitioner）。其中，哈希分区是根据哈希值进行分区，范围分区是将一定范围的数据映射到一个分区中。这两种分区方式已经可以满足大多数应用场景的需求。与此同时，Spark 也支持自定义分区方式，即通过一个自定义的 Partitioner 对象来控制 RDD 的分区，从而进一步减少通信开销。

3.3.3　只读

RDD 是只读的，要想改变 RDD 中的数据，只能在现有 RDD 基础上创建新的 RDD。

由一个 RDD 转换到另一个 RDD，可以通过丰富的转换算子实现，不再像 MapReduce 那样只能写 map 和 reduce 了。

3.3.4　依赖（血缘）

RDD 之间通过操作算子进行转换，转换得到的新 RDD 包含了从其他 RDD 衍生所必需的信息，RDD 之间维护着这种血缘关系，也称之为依赖。

RDD 之间的依赖包括两种窄依赖和宽依赖，如图 3-1 所示。

- 窄依赖，RDD 之间分区是一一对应的。

- 宽依赖，下游 RDD 的每个分区与上游 RDD（也称之为父 RDD）的每个分区都有关，是多对多的关系。

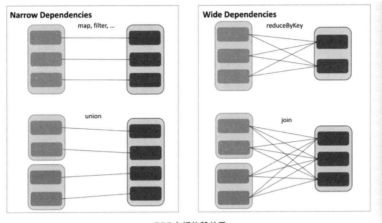

图 3-1

1）窄依赖

如果 B RDD 是由 A RDD 计算得到的，那么 B RDD 就是子 RDD，A RDD 就是父 RDD。

如果依赖关系在设计的时候就可以确定，而不需要考虑父 RDD 分区中的记录，并且父 RDD 中的每个分区最多只有一个子分区，那么这样的依赖就叫窄依赖，如图 3-2 所示。

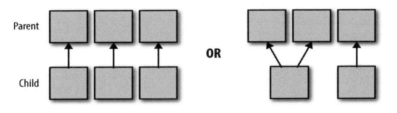

图 3-2

一句话总结：父 RDD 的每个分区最多被一个 RDD 的分区使用。

具体来说，窄依赖的时候，子 RDD 中的分区要么只依赖一个父 RDD 中的一个分区（比如 map、filter 操作），要么在设计时候就能确定子 RDD 是父 RDD 的一个子集（比如 coalesce 操作）。

所以，窄依赖的转换可以在任何一个分区上单独执行，而不需要其他分区的任何信息。

2）宽依赖

如果父 RDD 的分区被不止一个子 RDD 的分区依赖，那么就是宽依赖。宽依赖如图 3-3 所示。

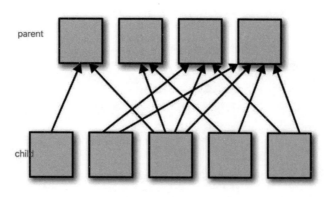

图 3-3

宽依赖工作的时候，不能随意在某些记录上运行，而是需要使用特殊的方式（比如按照 key）来获取分区中的所有数据。

例如：在排序（sort）的时候，数据必须被分区，同样范围的 key 必须在同一个分区内。

具有宽依赖的 transformation 包括：sort、reduceByKey、groupByKey、join 和调用 rePartition 函数的任何操作。

3.3.5 缓存

如果在应用程序中多次使用同一个 RDD，那么可以将该 RDD 缓存起来，该 RDD 只有在第一次计算的时候会根据依赖关系得到分区的数据，在后续其他地方用到该 RDD 的时候，会直接从缓存处获取而不用再根据依赖关系进行计算，这样就加速了后期的重用。

如图 3-4 所示，RDD-1 经过一系列的转换后得到 RDD-n 并保存到 HDFS，RDD-1 在这一过程中会有一个中间结果，如果将这个中间结果缓存到内存，那么在随后的 RDD-1 转换到 RDD-m 这一过程中，就不会计算其之前的 RDD-0 了。

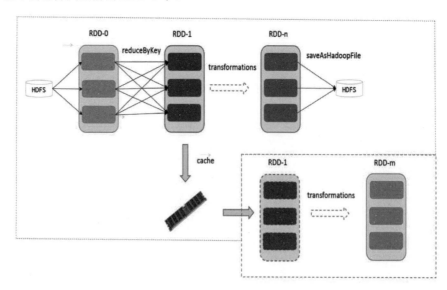

图 3-4

3.3.6 checkpoint

RDD 的依赖关系天然地可以实现容错，当 RDD 的某个分区数据计算失败或丢失时，可以通过依赖关系重建。但是，对于长时间迭代型应用来说，随着迭代的进行，RDD 之间的依赖关系会越来越长，一旦在后续迭代过程中出错，则需要通过非常长的依赖关系去重建，势必影响性能。为此，RDD 支持 checkpoint 将数据保存到持久化的存储中，这样就可以切断之前的依赖关系，因为 checkpoint 后的 RDD 不需要知道它的父 RDD 了，它可以直接从 checkpoint 处拿到数据。

3.4 RDD 的创建与处理过程

本节主要介绍 RDD 的创建及其处理过程。

3.4.1 RDD 的创建

Spark 可以从 Hadoop 支持的任何存储源中加载数据去创建 RDD，包括本地文件系统和 HDFS 等文件系统。我们通过 Spark 中的 SparkContext 对象调用 textFile()方法来加载数据创建 RDD。

（1）从文件系统加载数据创建 RDD：

```
scala> val test=sc.textFile("file:///export/data/test.txt")
test: org.apache.spark.rdd.RDD[String]=file:///export/data/test.txt
MapPartitionsRDD[1] at textFile at <console>:24
```

（2）从 HDFS 中加载数据创建 RDD：

```
scala> val testRDD=sc.textFile("/data/test.txt")
testRDD:org.apache.spark.rdd.RDD[String]=/data/test.txt MapPartitionsRDD[1]
at textFile at <console>:24
```

（3）Spark 还可以通过并行集合创建 RDD，即在一个已经存在的集合数组上，通过 SparkContext 对象调用 parallelize()方法来创建 RDD：

```
scala> val array=Array(1,2,3,4,5)
array: Array[Int]=Array(1,2,3,4,5)
scala> val arrRDD=sc.parallelize(array)
arrRDD: org.apache.spark.rdd.RDD[Int]=ParallelcollectionRDD[6] at parallelize
at <console>:26
```

3.4.2 RDD 的处理过程

Spark 用 Scala 语言实现了 RDD 的 API，程序开发者可以通过调用 API 对 RDD 进行操作。RDD 经过一系列的"转换"操作，每一次转换都会产生不同的 RDD，以供下一次"转换"操作使用，直到最后一个 RDD 经过"行动"操作才会被真正计算处理，并输出到外部数据源中，若是中间的数据结果需要复用，则可以进行缓存处理，将数据缓存到内存中。整个处理过程如图 3-5 所示。

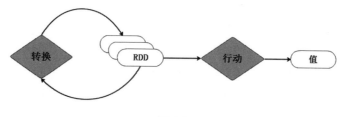

图 3-5

3.4.3 RDD 的算子

RDD 的操作算子包括两类，一类叫作 transformation（转换算子），它是用来将 RDD 进行转化，构建 RDD 的依赖关系；另一类叫作 action（行动算子），它是用来触发 RDD 的计算，得到 RDD 的相关计算结果或者将 RDD 保存的文件系统中。两类算子的区别如表 3-1 所示。

表3-1 两类算子的比较

算　子	算子函数例子	区　别
tansformation	map、filter、groupBy、join、union、reduce、sort、partitionBy	返回值还是 RDD，不会马上提交 Spark 集群运行
action	count、collect、take、save、show	返回值不是 RDD，会形成 DAG，提交 Spark 集群运行并立即返回结果

算子的功能主要包括：

- 通过转换算子，获取一个新的 RDD。
- 通过行动算子，触发 Spark Job 提交作业。

3.4.4 常见的转换算子

常见的转换算子（transformaction）可以分为两类：一类是 Value 数据类型的转换算子，这种变换不触发提交作业，针对处理的数据项是 Value 型的数据；另一类是 Key-Value 数据类型的转换算子，这种变换也不触发提交作业，针对处理的数据项是 Key-Value 型的数据。

1. Value 型转换算子

常见的 Value 型转换算子如下：

1）map

map 将数据集中的每个元素通过用户自定义函数转换成一个新的 RDD，新的 RDD 叫作 MappedRDD。示例如下：

```
val a = sc.parallelize(List("dog", "salmon", "salmon", "rat", "elephant"), 3)
val b = a.map(_.length)
val c = a.zip(b)
c.collect
```

上面示例中，zip 函数用于将两个 RDD 组合成 Key-Value 形式的 RDD。

结果：

```
res0: Array[(String, Int)] = Array((dog,3), (salmon,6), (salmon,6), (rat,3),
(elephant,8))
```

2）flatMap

flatMap 与 map 类似，但每个元素输入项都可以被映射到 0 个或多个的输出项，最终将结果"扁平化"后输出。示例如下：

```
val a = sc.parallelize(1 to 10, 5)
a.flatMap(1 to _).collect
```

结果：

```
res1: Array[Int] = Array(1, 1, 2, 1, 2, 3, 1, 2, 3, 4, 1, 2, 3, 4, 5, 1, 2, 3,
4, 5, 6, 1, 2, 3, 4, 5, 6, 7, 1, 2, 3, 4, 5, 6, 7, 8, 1, 2, 3, 4, 5, 6, 7, 8, 9,
1, 2, 3, 4, 5, 6, 7, 8, 9, 10)
```

又如：

```
sc.parallelize(List(1, 2, 3), 2).flatMap(x => List(x, x, x)).collect
```

结果：

```
res2: Array[Int] = Array(1, 1, 1, 2, 2, 2, 3, 3, 3)
```

3）mapPartitions

mapPartitions 类似于 map，但 map 作用于每个分区的每个元素，而 mapPartitions 作用于每个分区的 func 的类型：Iterator[T]=>Iterator[U]。假设有 N 个元素，有 M 个分区，那么 map 的函数将被调用 N 次，而 mapPartitions 被调用 M 次。当在映射的过程中不断地创建对象时就可以使用 mapPartitions，它比 map 的效率要高很多。比如向数据库写入数据时，如果使用 map，就需要为每个元素创建 connection 对象，但使用 mapPartitions 的话，就只需要为每个分区创建 connection 对象。示例如下：

```
val l = List(("kpop","female"),("zorro","male"),("mobin","male"), ("lucy","female"))
val rdd = sc.parallelize(l,2)
rdd.mapPartitions(x => x.filter(_._2 == "female")).foreachPartition(p=>{
    println(p.toList)
    println("====分区分割线====" )
})
```

结果：

```
====分区分割线====
List((kpop,female))
====分区分割线====
List((lucy,female))
```

4）glom

glom 将 RDD 每个分区中类型为 T 的元素转换为数组 Array[T]。示例如下：

```
val a = sc.parallelize(1 to 100, 3)
```

```
a.glom.collect
```

结果:

```
res3: Array[Array[Int]] = Array(Array(1, 2, 3, 4, 5, 6, 7, 8, 9, 10, 11, 12,
13, 14, 15, 16, 17, 18, 19, 20, 21, 22, 23, 24, 25, 26, 27, 28, 29, 30, 31, 32, 33),
Array(34, 35, 36, 37, 38, 39, 40, 41, 42, 43, 44, 45, 46, 47, 48, 49, 50, 51, 52,
53, 54, 55, 56, 57, 58, 59, 60, 61, 62, 63, 64, 65, 66), Array(67, 68, 69, 70, 71,
72, 73, 74, 75, 76, 77, 78, 79, 80, 81, 82, 83, 84, 85, 86, 87, 88, 89, 90, 91, 92,
93, 94, 95, 96, 97, 98, 99, 100))
```

5) union

union 将两个 RDD 中的数据集进行合并,最终返回两个 RDD 的并集。若 RDD 中存在相同的元素,也不会去重。示例如下:

```
val a = sc.parallelize(1 to 3, 1)
val b = sc.parallelize(1 to 7, 1)
a.union(b).collect
```

结果:

```
res4: Array[Int] = Array(1, 2, 3, 5, 6, 7)
```

6) cartesian

cartesian 对两个 RDD 中的所有元素进行笛卡尔积操作。示例如下:

```
val x = sc.parallelize(List(1,2,3,4,5))
val y = sc.parallelize(List(6,7,8,9,10))
x.cartesian(y).collect
```

结果:

```
res5: Array[(Int, Int)] = Array((1,6), (1,7), (1,8), (1,9), (1,10), (2,6), (2,7),
(2,8), (2,9), (2,10), (3,6), (3,7), (3,8), (3,9), (3,10), (4,6), (5,6), (4,7), (5,7),
(4,8), (5,8), (4,9), (4,10), (5,9), (5,10))
```

7) groupBy

groupBy 用于将 RDD 中的元素按照自定义规则进行数据分组,相同 key 的数据放在一起。示例如下:

```
val a = sc.parallelize(1 to 9, 3)
a.groupBy(x => { if (x % 2 == 0) "even" else "odd" }).collect
```

结果:

```
res6: Array[(String, Seq[Int])] = Array((even,ArrayBuffer(2, 4, 6, 8)),
(odd,ArrayBuffer(1, 3, 5, 7, 9)))
```

8) filter

filter 对元素进行过滤,对每个元素应用指定函数,返回值为 true 的元素将保留在 RDD 中,返回值为 false 的元素将被过滤掉。示例如下:

```
val a = sc.parallelize(1 to 10, 3)
val b = a.filter(_ % 2 == 0)
b.collect
```

结果：

```
res7: Array[Int] = Array(2, 4, 6, 8, 10)
```

9）distinct

distinct 用于去重。示例如下：

```
val c = sc.parallelize(List("Gnu", "Cat", "Rat", "Dog", "Gnu", "Rat"), 2)
c.distinct.collect
```

结果：

```
res8: Array[String] = Array(Dog, Gnu, Cat, Rat)
```

10）subtract

subtract 去掉含有重复的项。示例如下：

```
val a = sc.parallelize(1 to 9, 3)
val b = sc.parallelize(1 to 3, 3)
val c = a.subtract(b)
c.collect
```

结果：

```
res9: Array[Int] = Array(6, 9, 4, 7, 5, 8)
```

2. Key-Value 型转换算子

常用的 Key-Value 型转换算子如下：

1）mapValues

mapValues 是针对[K,V]中的 V 值进行 map 操作。示例如下：

```
val a = sc.parallelize(List("dog", "tiger", "lion", "cat", "panther", "eagle"), 2)
val b = a.map(x => (x.length, x))
b.mapValues("x" + _ + "x").collect
```

结果：

```
res14: Array[(Int, String)] = Array((3,xdogx), (5,xtigerx), (4,xlionx), (3,xcatx), (7,xpantherx), (5,xeaglex))
```

2）combineByKey

combineByKey 使用用户设置好的聚合函数对每个 Key 中的 Value 进行组合，可以将输入类型由 RDD[(K, V)]转换成 RDD[(K, C)]。示例如下：

```
val a = sc.parallelize(List("dog","cat","gnu","salmon","rabbit","turkey","wolf","bear","bee"), 3)
```

```
    val b = sc.parallelize(List(1,1,2,2,2,1,2,2,2), 3)
    val c = b.zip(a)
    val d = c.combineByKey(List(_), (x:List[String], y:String) => y :: x,
(x:List[String], y:List[String]) => x ::: y)
    d.collect
```

结果：

```
    res15: Array[(Int, List[String])] = Array((1,List(cat, dog, turkey)),
(2,List(gnu, rabbit, salmon, bee, bear, wolf)))
```

3）reduceByKey

reduceByKey 是对元素为键-值对的 RDD 中 Key 相同的元素的 Value 进行 binary_function 的 reduce 操作键-值，因此 Key 相同的多个元素的值被 reduce 为一个值，然后与原 RDD 中的 Key 组成一个新的键-值对。示例如下：

```
    val a = sc.parallelize(List("dog", "cat", "owl", "gnu", "ant"), 2)
    val b = a.map(x => (x.length, x))
    b.reduceByKey(_ + _).collect
```

结果：

```
    res16: Array[(Int, String)] = Array((3,dogcatowlgnuant))
```

又如：

```
    val a = sc.parallelize(List("dog", "tiger", "lion", "cat", "panther", "eagle"),
2)
    val b = a.map(x => (x.length, x))
    b.reduceByKey(_ + _).collect
```

结果：

```
    res17: Array[(Int, String)] = Array((4,lion), (3,dogcat), (7,panther),
(5,tigereagle))
```

4）partitionBy

对 RDD 进行分区操作。示例如下：

```
    val rdd1 = sc.makeRDD(Array((1,"a"),(1,"b"),(2,"b"),(3,"c"),(4,"d")),4)
    rdd1.partitionBy(new org.apache.spark.HashPartitioner(2)).glom.collect
```

结果：

```
    Array[Array[(Int, String)]] = Array(Array((2,b), (4,d)), Array((1,a), (1,b),
(3,c)))
```

5）cogroup

cogroup 指对两个元素为键-值对的 RDD，将每个 RDD 中相同 key 中的元素分别聚合成一个集合。示例如下：

```
    val a = sc.parallelize(List(1, 2, 1, 3), 1)
```

```
val b = a.map((_, "b"))
val c = a.map((_, "c"))
b.cogroup(c).collect
```

结果：

```
res18:Array[(Int, (Iterable[String], Iterable[String]))] = Array(
(2,(ArrayBuffer(b),ArrayBuffer(c))),
(3,(ArrayBuffer(b),ArrayBuffer(c))),
(1,(ArrayBuffer(b, b),ArrayBuffer(c, c))))
```

6）join

join 是对两个需要连接的 RDD 进行 cogroup 函数操作。示例如下：

```
val a = sc.parallelize(List("dog", "salmon", "salmon", "rat",
"elephant"), 3)
val b = a.keyBy(_.length)
val c = sc.parallelize(List("dog","cat","gnu","salmon","rabbit","turkey",
"wolf","bear","bee"), 3)
val d = c.keyBy(_.length)
b.join(d).collect
```

结果：

```
res19: Array[(Int, (String, String))] = Array((6,(salmon,salmon)),
(6,(salmon,rabbit)), (6,(salmon,turkey)), (6,(salmon,salmon)),
(6,(salmon,rabbit)), (6,(salmon,turkey)), (3,(dog,dog)), (3,(dog,cat)),
(3,(dog,gnu)), (3,(dog,bee)), (3,(rat,dog)), (3,(rat,cat)), (3,(rat,gnu)),
(3,(rat,bee)))
```

3.4.5 常见的行动算子

行动算子（action）会触发 SparkContext 提交作业，常见的行动算子如下：

1. foreach

foreach 用于打印输出。示例如下：

```
val c = sc.parallelize(List("cat", "dog", "tiger", "lion", "gnu", "crocodile",
"ant", "whale", "dolphin", "spider"), 3)
c.foreach(x => println(x + "s are yummy"))
```

结果：

```
lions are yummy
gnus are yummy
crocodiles are yummy
ants are yummy
whales are yummy
dolphins are yummy
spiders are yummy
```

2. saveAsTextFile

saveAsTextFile 用于保存结果到 HDFS。示例如下：

```
val a = sc.parallelize(1 to 10000, 3)
a.saveAsTextFile("/user/yuhui/mydata_a")
```

结果：

```
[root@tagtic-slave03 ~]# Hadoop fs -ls /user/yuhui/mydata_a
Found 4 items
-rw-r-r- 2 root supergroup 0 2017-05-22 14:28 /user/yuhui/mydata_a/_SUCCESS
-rw-r-r- 2 root supergroup 15558 2017-05-22 14:28 /user/yuhui/mydata_a/part-00000
-rw-r-r- 2 root supergroup 16665 2017-05-22 14:28 /user/yuhui/mydata_a/part-00001
-rw-r-r- 2 root supergroup 16671 2017-05-22 14:28 /user/yuhui/mydata_a/part-00002
```

3. saveAsObjectFile

saveAsObjectFile 用于将 RDD 中的元素序列化成对象，并存储到文件中。对于 HDFS，默认采用 SequenceFile 保存。示例如下：

```
val x = sc.parallelize(1 to 100, 3)
x.saveAsObjectFile("/user/yuhui/objFile")
val y = sc.objectFile[Int]("/user/yuhui/objFile")
y.collect
```

结果：

```
res22: Array[Int] = Array[Int] = Array(1, 2, 3, 4, 5, 6, 7, 8, 9, 10, 11, 12, 13, 14, 15, 16, 17, 18, 19, 20, 21, 22, 23, 24, 25, 26, 27, 28, 29, 30, 31, 32, 33, 34, 35, 36, 37, 38, 39, 40, 41, 42, 43, 44, 45, 46, 47, 48, 49, 50, 51, 52, 53, 54, 55, 56, 57, 58, 59, 60, 61, 62, 63, 64, 65, 66, 67, 68, 69, 70, 71, 72, 73, 74, 75, 76, 77, 78, 79, 80, 81, 82, 83, 84, 85, 86, 87, 88, 89, 90, 91, 92, 93, 94, 95, 96, 97, 98, 99, 100)
```

4. collect

collect 用于将 RDD 中的数据收集起来，变成一个数组，仅限数据量比较小的时候使用。示例如下：

```
val c = sc.parallelize(List("Gnu", "Cat", "Rat", "Dog", "Gnu", "Rat"), 2)
c.collect
```

结果：

```
res23: Array[String] = Array(Gnu, Cat, Rat, Dog, Gnu, Rat)
```

5. collectAsMap

collectAsMap 用于返回 hashMap，包含所有 RDD 中的分片，key 如果重复，那后边的元素会覆盖前面的元素。示例如下：

```
val a = sc.parallelize(List(1, 2, 1, 3), 1)
val b = a.zip(a)
b.collectAsMap
```

上面示例中，zip 函数用于将两个 RDD 组合成键-值形式的 RDD。

结果：

```
res24: Scala.collection.Map[Int,Int] = Map(2 -> 2, 1 -> 1, 3 -> 3)
```

6. reduceByKeyLocally

reduceByKeyLocally 先执行 reduce，然后再执行 collectAsMap。示例如下：

```
val a = sc.parallelize(List("dog", "cat", "owl", "gnu", "ant"), 2)
val b = a.map(x => (x.length, x))
b.reduceByKey(_ + _).collect
```

结果：

```
res25: Array[(Int, String)] = Array((3,dogcatowlgnuant))
```

7. lookup

lookup 用于针对 key-value 类型的 RDD 进行查找。示例如下：

```
val a = sc.parallelize(List("dog", "tiger", "lion", "cat", "panther", "eagle"), 2)
val b = a.map(x => (x.length, x))
b.lookup(3)
```

结果：

```
res26: Seq[String] = WrappedArray(tiger, eagle)
```

8. count

count 用于计算总数。示例如下：

```
val c = sc.parallelize(List("Gnu", "Cat", "Rat", "Dog"), 2)
c.count
```

结果：

```
res27: Long = 4
```

9. top(k)

top 用于返回最大的 k 个元素。示例如下：

```
val c = sc.parallelize(Array(6, 9, 4, 7, 5, 8), 2)
```

```
c.top(2)
```

结果:

```
res28: Array[Int] = Array(9, 8)
```

10. reduce

reduce 相当于对 RDD 中的元素进行 reduceLeft 函数的操作。示例如下:

```
val a = sc.parallelize(1 to 100, 3)
a.reduce(_ + _)
```

结果:

```
res29: Int = 5050
```

第 4 章

Spark SQL 结构化数据文件处理

在很多情况下,开发人员并不了解 Scala 语言,也不了解 Spark 常用的 API,但又非常想要使用 Spark 框架提供的强大的数据分析能力。Spark 的开发工程师们考虑到了这个问题,于是利用 SQL 语言的语法简洁、学习门槛低以及在编程语言中普及程度和流行程度高等诸多优势,开发了 Spark SQL 模块。通过 Spark SQL,开发人员能够使用 SQL 语句实现对结构化数据的处理。

本章主要知识点:

- Spark SQL 概述
- Spark SQL 编程
- Spark SQL 数据源

4.1 Spark SQL 概述

本节主要介绍什么是 Spark SQL、Spark SQL 的特点,以及 DataFrame 和 DataSet 的相关内容。

4.1.1 什么是 Spark SQL

Spark SQL 是 Spark 用于结构化数据(Structured Data)处理的 Spark 模块。与基本的 Spark RDD API 不同,Spark SQL 的抽象数据类型为 Spark 提供了关于数据结构和正在执行的计算的更多信息。在内部,Spark SQL 使用这些额外的信息去做一些优化。

有多种方式与 Spark SQL 进行交互,比如 SQL 和 Dataset API。当计算结果的时候,这些接口使用相同的执行引擎,不依赖正在使用哪种 API 或者语言。这种统一也就意味着开发者可以很容易在不同的 API 之间进行切换,这些 API 提供了最自然的方式来表达给定的转换。

Hive 是将 Hive SQL 转换成 MapReduce,然后提交到集群上执行,大大简化了编写 MapReduce 程序的复杂性,但是 MapReduce 这种计算模型执行效率比较慢,所以 Spark SQL 应运而生,它将 Spark SQL 转换成 RDD,然后提交到集群上执行,执行效率非常高。

Spark SQL 提供了以下 2 个编程抽象,类似 Spark Core 中的 RDD。

- DataFrame。

- DataSet。

4.1.2 Spark SQL 的特点

1. Integrated（易整合）

Spark SQL 无缝地整合了 SQL 查询和 Spark 编程，如图 4-1 所示。

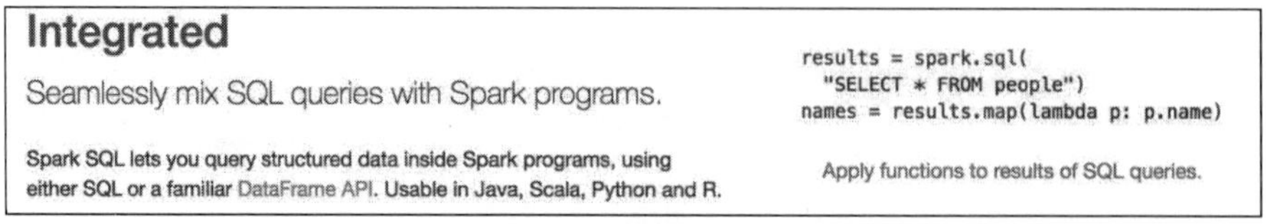

图 4-1

2. Uniform Data Access（统一的数据访问方式）

Spark SQL 使用相同的方式连接不同的数据源，如图 4-2 所示。

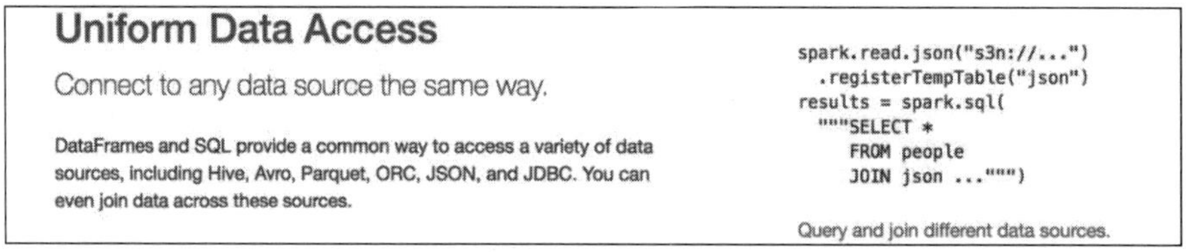

图 4-2

3. Hive Integration（集成 Hive）

Spark SQL 在已有的仓库上直接运行 SQL 或者 HiveQL，如图 4-3 所示。

图 4-3

4. Standard Connectivity（标准的连接方式）

Spark SQL 通过 JDBC 或者 ODBC 来连接，如图 4-4 所示。

图 4-4

4.1.3 什么是 DataFrame

与 RDD 类似，DataFrame 也是一个分布式数据容器。然而 DataFrame 更像传统数据库的二维表格，除了数据以外，还记录数据的结构信息，即 schema。

同时，DataFrame 与 Hive 类似，也支持嵌套数据类型（struct、array 和 map）。

从 API 易用性的角度上看，DataFrame API 提供的是一套高层的关系操作，比函数式的 RDD API 要更加友好、门槛更低。

DataFrame 与 RDD 的区别如图 4-5 所示。

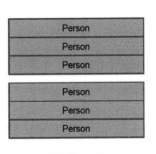

图 4-5

图中左侧的 RDD[Person]虽然以 Person 为类型参数，但 Spark 框架本身不了解 Person 类的内部结构。而右侧的 DataFrame 却提供了详细的结构信息，使得 Spark SQL 可以清楚地知道该数据集中包含哪些列，每列的名称和类型各是什么。

DataFrame 为数据提供了 schema 视图，可以把它当作数据库中的一张表来对待。

DataFrame 也是懒执行的，性能上比 RDD 要高，主要原因在于优化的执行计划——查询计划通过 Spark Catalyst Optimizer（Catalyst 优化器，基于 Scala 的函数式编程结构设计的可扩展优化器）进行优化。

为了说明查询优化，我们来看图 4-6 展示的人口数据分析的示例。

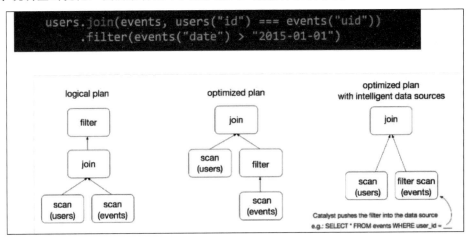

图 4-6

图中构造了两个 DataFrame，将它们 join 之后又做了一次 filter 操作。

如果原封不动地执行这个计划，最终的执行效率不是很高，因为 join 是一个代价较大的操作，也可能会产生一个较大的数据集。如果我们能将 filter 下推到 join 下方，先对 DataFrame 进行过滤，再 join 过滤后的较小的结果集，便可以有效缩短执行时间。而 Spark SQL 的查询优化器正是这样做的。简而言之，逻辑查询计划优化就是一个利用基于关系代数的等价变换，将高成本操作替换为低成本操作的过程。

4.1.4　什么是 DataSet

（1）DataSet 是 DataFrame API 的一个扩展，也是 Spark SQL 最新的数据抽象（1.6 版本新增）。

（2）用户友好的 API 风格，既具有类型安全检查，也具有 DataFrame 的查询优化特性。

（3）Dataset 支持编解码器，当需要访问非堆上的数据时可以避免反序列化整个对象，提高了效率。

（4）样例类被用来在 DataSet 中定义数据的结构信息，样例类中每个属性的名称直接映射到 DataSet 中的字段名称。

（5）DataFrame 是 DataSet 的特例，DataFrame=DataSet[Row]，因此可以通过 as 方法将 DataFrame 转换为 DataSet。Row 是一个类型，跟 Car、Person 这些类型一样，所有的表结构信息都用 Row 来表示。

（6）DataSet 是强类型的，比如可以有 DataSet[Car]、DataSet[Person]等。

（7）DataFrame 只是知道字段，但是不知道字段的类型，所以在执行这些操作时是没办法在编译的时候检查字段类型是否正确，比如我们对一个字符串进行减法操作，在执行的时候才报错。而 DataSet 不仅知道字段，还知道字段类型，所以有更严格的错误检查。

4.2　Spark SQL 编程

本节主要介绍如何使用 DataFrame 和 DataSet 进行编程，以及 DataFrame 和 DataSet 之间的关系和转换方法。关于具体的 SQL 书写不是本节的重点。

4.2.1　SparkSession

在旧版本中，Spark SQL 提供两种 SQL 查询起始点：一个叫作 SQLContext，用于 Spark 自己提供的 SQL 查询；一个叫作 HiveContext，用于连接 Hive 的查询。

SparkSession 是 Spark 最新的 SQL 查询起始点，实质上是 SQLContext 和 HiveContext 的组合，所以在 SQLContext 和 HiveContext 上可用的 API 在 SparkSession 上同样是可以使用的。

SparkSession 内部封装了 SparkContext，所以计算实际上是由 SparkContext 完成的。

当我们使用 spark-shell 的时候，Spark 会自动创建一个叫作 spark 的 SparkSession，就像以前可以自动获取到一个 sc 来表示 SparkContext，如图 4-7 所示。

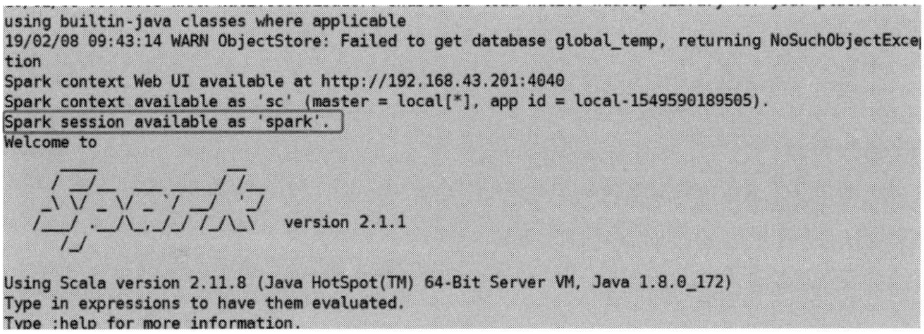

```
using builtin-java classes where applicable
19/02/08 09:43:14 WARN ObjectStore: Failed to get database global_temp, returning NoSuchObjectExce
tion
Spark context Web UI available at http://192.168.43.201:4040
Spark context available as 'sc' (master = local[*], app id = local-1549590189505).
Spark session available as 'spark'.
Welcome to
      ____              __
     / __/__  ___ _____/ /__
    _\ \/ _ \/ _ `/ __/  '_/
   /___/ .__/\_,_/_/ /_/\_\   version 2.1.1
      /_/

Using Scala version 2.11.8 (Java HotSpot(TM) 64-Bit Server VM, Java 1.8.0_172)
Type in expressions to have them evaluated.
Type :help for more information.
```

图 4-7

4.2.2 使用 DataFrame 进行编程

Spark SQL 的 DataFrame API 允许我们使用 DataFrame 而不用必须去注册临时表或者生成 SQL 表达式。DataFrame API 既有 transformation 操作也有 action 操作，DataFrame 的转换从本质上来说更具有关系，而 DataSet API 提供了更加函数式的 API。

1. 创建 DataFrame

有了 SparkSession 之后，可以通过 SparkSession 的以下 3 种方式来创建 DataFrame：

- 通过 Spark 的数据源来创建。
- 通过已知的 RDD 来创建。
- 通过查询一个 Hive 表来创建。

Spark 支持的数据源如下：

```
scala> spark.read.
csv      jdbc    load     options   parquet  table   textFile
format   json    option   orc       schema   text
```

通过 Spark 数据源创建 DataFrame 的代码如下：

```
// 读取 JSON 文件
scala> val df = spark.read.json("/opt/module/spark-local/examples/src/main/resources/employees.json")
df: org.apache.spark.sql.DataFrame = [name: string, salary: bigint]

// 展示结果
scala> df.show
+-------+------+
|   name|salary|
+-------+------+
|Michael|  3000|
|   Andy|  4500|
| Justin|  3500|
|  Berta|  4000|
+-------+------+
```

利用其他数据源创建 DataFrame 将在 4.3.3 节和 4.3.4 节介绍。

2. DataFrame 语法风格

1）SQL 语法风格

SQL 语法风格是指我们查询数据的时候使用 SQL 语句。这种风格的查询必须有临时视图或者全局视图来辅助。以下是创建临时视图的代码：

```
scala> val df = 
spark.read.json("/opt/module/spark-local/examples/src/main/resources/people.json")
   df: org.apache.spark.sql.DataFrame = [age: bigint, name: string]

   scala> df.createOrReplaceTempView("people")

   scala> spark.sql("select * from people").show
   +----+-------+
   | age|   name|
   +----+-------+
   |null|Michael|
   |  30|   Andy|
   |  19| Justin|
   +----+-------+
```

注意：（1）临时视图只能在当前 Session 中有效，在新的 Session 中无效。

（2）可以创建全局视图。访问全局视图需要全路径，如 global_temp.xxx。

以下是创建全局视图的代码：

```
scala> val df = 
spark.read.json("/opt/module/spark-local/examples/src/main/resources/people.json")
   df: org.apache.spark.sql.DataFrame = [age: bigint, name: string]

   scala> df.createGlobalTempView("people")

   scala> spark.sql("select * from global_temp.people")
   res31: org.apache.spark.sql.DataFrame = [age: bigint, name: string]

   scala> res31.show
   +----+-------+
   | age|   name|
   +----+-------+
   |null|Michael|
   |  30|   Andy|
   |  19| Justin|
   +----+-------+
```

2）DSL 语法风格

DataFrame 提供一个特定领域语言（Domain-Specific Language，DSL）去管理结构化的数据。

可以在 Scala、Java、Python 和 R 中使用 DSL。使用 DSL 语法风格不必去创建临时视图了。

（1）查看 schema 信息，示例代码如下：

```
scala> val df =
spark.read.json("/opt/module/spark-local/examples/src/main/resources/people.json")
df: org.apache.spark.sql.DataFrame = [age: bigint, name: string]

scala> df.printSchema
root
 |-- age: long (nullable = true)
 |-- name: string (nullable = true)
```

（2）使用 DSL 查询，示例代码如下：

只查询 name 列数据：

```
scala> df.select($"name").show
+-------+
|   name|
+-------+
|Michael|
|   Andy|
| Justin|
+-------+


scala> df.select("name").show
+-------+
|   name|
+-------+
|Michael|
|   Andy|
| Justin|
+-------+
```

查询 name 和 age：

```
scala> df.select("name", "age").show
+-------+----+
|   name| age|
+-------+----+
|Michael|null|
|   Andy|  30|
| Justin|  19|
+-------+----+
```

查询 name 和 age + 1：

```
scala> df.select($"name", $"age" + 1).show
+-------+---------+
|   name|(age + 1)|
+-------+---------+
```

```
|Michael|    null|
|   Andy|      31|
| Justin|      20|
+-------+--------+
```

注意：涉及运算的时候，每列都必须使用 $。

查询 age 大于 20 的数据：

```
scala> df.filter($"age" > 21).show
+---+----+
|age|name|
+---+----+
| 30|Andy|
+---+----+
```

按照 age 分组，查看数据条数：

```
scala> df.groupBy("age").count.show
+----+-----+
| age|count|
+----+-----+
|  19|    1|
|null|    1|
|  30|    1|
+----+-----+
```

3. RDD 和 DataFrame 的交互

1）从 RDD 到 DataFrame

涉及 RDD、DataFrame、DataSet 之间的操作时，需要进行导入，即 import spark.implicits._。这里的 spark 不是包名，而是表示 SparkSession 的那个对象，所以必须先创建 SparkSession 对象再导入。implicits 是一个内部 object。

首先创建一个 RDD：

```
scala> val rdd1 = 
sc.textFile("/opt/module/spark-local/examples/src/main/resources/people.txt")
    rdd1: org.apache.spark.rdd.RDD[String] = 
/opt/module/spark-local/examples/src/main/resources/people.txt 
MapPartitionsRDD[10] at textFile at <console>:24
```

然后进行转换，转换 3 种方法：手动转换、通过样例类反身转换和通过 API 的方式转换。

（1）手动转换。

示例代码如下：

```
scala> val rdd2 = rdd1.map(line => { val paras = line.split(", "); (paras(0), 
paras(1).toInt) })
    rdd2: org.apache.spark.rdd.RDD[(String, Int)] = MapPartitionsRDD[11] at map at 
<console>:26

    // 转换为 DataFrame 的时候手动指定每个数据字段名
```

```
scala> rdd2.toDF("name", "age").show
+-------+---+
|   name|age|
+-------+---+
|Michael| 29|
|   Andy| 30|
| Justin| 19|
+-------+---+
```

（2）通过样例类反射转换。

首先创建样例类：

```
scala> case class People(name :String, age: Int)
defined class People
```

然后使用样例把 RDD 转换成 DataFrame：

```
scala> val rdd2 = rdd1.map(line => { val paras = line.split(", "); People(paras(0), paras(1).toInt) })
rdd2: org.apache.spark.rdd.RDD[People] = MapPartitionsRDD[6] at map at <console>:28
```

```
scala> rdd2.toDF.show
+-------+---+
|   name|age|
+-------+---+
|Michael| 29|
|   Andy| 30|
| Justin| 19|
+-------+---+
```

（3）通过 API 的方式转换。

示例代码如下：

代码 4-1　DataFrameDemo2.scala

```scala
import org.apache.spark.SparkContext
import org.apache.spark.rdd.RDD
import org.apache.spark.sql.types.{IntegerType, StringType, StructField, StructType}
import org.apache.spark.sql.{DataFrame, Dataset, Row, SparkSession}

object DataFrameDemo2 {
    def main(args: Array[String]): Unit = {
        val spark: SparkSession = SparkSession.builder()
            .master("local[*]")
            .appName("Word Count")
            .getOrCreate()
        val sc: SparkContext = spark.sparkContext
        val rdd: RDD[(String, Int)] = sc.parallelize(Array(("lisi", 10), ("zs", 20), ("zhiling", 40)))
        // 映射出来一个 RDD[Row]，因为 DataFrame 其实就是 DataSet[Row]
```

```
            val rowRdd: RDD[Row] = rdd.map(x => Row(x._1, x._2))
            // 创建 StructType 类型
            val types = StructType(Array(StructField("name", StringType),
StructField("age", IntegerType)))
            val df: DataFrame = spark.createDataFrame(rowRdd, types)
            df.show
    }
}
```

2）从 DataFrame 到 RDD

直接调用 DataFrame 的 rdd 方法就完成了转换。示例代码如下：

```
    scala> val df =
spark.read.json("/opt/module/spark-local/examples/src/main/resources/people.jso
n")
    df: org.apache.spark.sql.DataFrame = [age: bigint, name: string]

    scala> val rdd = df.rdd
    rdd: org.apache.spark.rdd.RDD[org.apache.spark.sql.Row] = MapPartitionsRDD[6]
at rdd at <console>:25

    scala> rdd.collect
    res0: Array[org.apache.spark.sql.Row] = Array([null,Michael], [30,Andy],
[19,Justin])
```

说明：得到的 RDD 中存储的数据类型是 org.apache.spark.sql.Row。

4.2.3 使用 DataSet 进行编程

DataSet 和 RDD 类似，但是 DataSet 没有使用 Java 序列化或者 Kryo 序列化，而是使用一种专门的编码器去序列化对象，然后在网络上处理或者传输。虽然编码器和标准序列化都负责将对象转换成字节，但编码器是动态生成的代码，使用的格式允许 Spark 执行许多操作，如过滤、排序和哈希，而无须将字节反序列化回对象。

DataSet 是具有强类型的数据集合，需要提供对应的类型信息。

1．创建 DataSet

（1）使用样例类的序列得到 DataSet。

示例代码如下：

```
    scala> case class Person(name: String, age: Int)
    defined class Person
    // 为样例类创建一个编码器
    scala> val ds = Seq(Person("lisi", 20), Person("zs", 21)).toDS
    ds: org.apache.spark.sql.Dataset[Person] = [name: string, age: int]
    scala> ds.show
    +----+---+
    |name|age|
    +----+---+
```

```
|lisi| 20|
|  zs| 21|
+----+---+
```

（2）使用基本类型的序列得到 DataSet。

示例代码如下：

```
// 基本类型的编码被自动创建
importing spark.implicits._
scala> val ds = Seq(1,2,3,4,5,6).toDS
ds: org.apache.spark.sql.Dataset[Int] = [value: int]
scala> ds.show
+-----+
|value|
+-----+
|    1|
|    2|
|    3|
|    4|
|    5|
|    6|
+-----+
```

说明：在实际使用的时候，很少把序列转换成 DataSet，更多的是通过 RDD 来得到 DataSet。

2. RDD 和 DataSet 的交互

1）从 RDD 到 DataSet

使用反射来推断包含特定类型对象的 RDD 的 schema。这种基于反射的方法可以生成更简洁的代码，并且当我们在编写 Spark 应用程序时已经知道模式，这种方法可以很好地工作。

为 Spark SQL 设计的 Scala API 可以自动地把包含样例类的 RDD 转换成 DataSet。样例类定义了表结构，样例类参数名通过反射被读取到，然后成为列名。样例类可以被嵌套，也可以包含复杂类型，像 Seq 或者 Array。

示例代码如下：

```
scala> val peopleRDD = sc.textFile("examples/src/main/resources/ people.txt")
peopleRDD: org.apache.spark.rdd.RDD[String] = 
examples/src/main/resources/people.txt MapPartitionsRDD[1] at textFile at 
<console>:24

scala> case class Person(name: String, age: Long)
defined class Person

scala> peopleRDD.map(line => {val para = 
line.split(",");Person(para(0),para(1).trim.toInt)}).toDS
res0: org.apache.spark.sql.Dataset[Person] = [name: string, age: bigint]
```

2）从 DataSet 到 RDD

将 DataSet 转换为 RDD 调用 rdd 方法即可。示例代码如下：

```
scala> val ds = Seq(Person("lisi", 40), Person("zs", 20)).toDS
ds: org.apache.spark.sql.Dataset[Person] = [name: string, age: bigint]

// 把 DataSet 转换成 RDD
scala> val rdd = ds.rdd
rdd: org.apache.spark.rdd.RDD[Person] = MapPartitionsRDD[8] at rdd at <console>:27

scala> rdd.collect
res5: Array[Person] = Array(Person(lisi,40), Person(zs,20))
```

4.2.4 DataFrame 和 DataSet 之间的交互

1. 从 DataFrame 到 DataSet

示例代码如下:

```
scala> val df = spark.read.json("examples/src/main/resources/people.json")
df: org.apache.spark.sql.DataFrame = [age: bigint, name: string]

scala> case class People(name: String, age: Long)
defined class People

// 将 DataFrame 转换成 DataSet
scala> val ds = df.as[People]
ds: org.apache.spark.sql.Dataset[People] = [age: bigint, name: string]
```

2. 从 DataSet 到 DataFrame

示例代码如下:

```
scala> case class Person(name: String, age: Long)
defined class Person

scala> val ds = Seq(Person("Andy", 32)).toDS()
ds: org.apache.spark.sql.Dataset[Person] = [name: string, age: bigint]

scala> val df = ds.toDF
df: org.apache.spark.sql.DataFrame = [name: string, age: bigint]

scala> df.show
+----+---+
|name|age|
+----+---+
|Andy| 32|
+----+---+
```

4.2.5 使用 IDEA 创建 Spark SQL 程序

使用 IDEA 创建 Spark SQL 程序,首先需要添加 Spark SQL 依赖:

```xml
<dependency>
    <groupId>org.apache.spark</groupId>
    <artifactId>spark-sql_2.11</artifactId>
    <version>2.1.1</version>
</dependency>
```

下面创建 Spark SQL 程序，代码如下：

代码 4-2　DataFrameDemo.scala

```scala
object DataFrameDemo {
    def main(args: Array[String]): Unit = {
        // 创建一个新的 SparkSession 对象
        val spark: SparkSession = SparkSession.builder()
            .master("local[*]")
            .appName("Word Count")
            .getOrCreate()
        // 导入用到隐式转换。若想要使用:$"age"，则必须导入
        val df = spark.read.json("file://" +
ClassLoader.getSystemResource("user.json").getPath)
        // 打印信息
        df.show
        // 查找年龄大于 19 岁的
        df.filter($"age" > 19).show

        // 创建临时表
        df.createTempView("user")
        spark.sql("select * from user where age > 19").show

        //关闭连接
        spark.stop()
    }
}
```

4.2.6　自定义 Spark SQL 函数

在 Shell 窗口中，用户可以通过 spark.udf 功能自定义函数。自定义 udf 函数的代码如下：

```
scala> val df = spark.read.json("examples/src/main/resources/people.json")
df: org.apache.spark.sql.DataFrame = [age: bigint, name: string]

scala> df.show
+----+-------+
| age|   name|
+----+-------+
|null|Michael|
|  30|   Andy|
|  19| Justin|
+----+-------+

// 注册一个 udf 函数: toUpper 是函数名，第二个参数是函数的具体实现
scala> spark.udf.register("toUpper", (s: String) => s.toUpperCase)
```

```
res1: org.apache.spark.sql.expressions.UserDefinedFunction =
UserDefinedFunction(<function1>,StringType,Some(List(StringType)))

scala> df.createOrReplaceTempView("people")

scala> spark.sql("select toUpper(name), age from people").show
+----------------+----+
|UDF:toUpper(name)| age|
+----------------+----+
|         MICHAEL|null|
|            ANDY|  30|
|           JUSTIN|  19|
+----------------+----+
```

4.3 Spark SQL 数据源

本节介绍 Spark SQL 的各种数据源。Spark SQL 的 DataFrame 接口支持操作多种数据源。一个 DataFrame 类型的对象可以像 RDD 那样操作（比如各种转换），也可以用来创建临时表。把 DataFrame 注册为一个临时表之后，就可以在它的数据上面执行 SQL 查询。

4.3.1 通用加载和保存函数

默认数据源是 parquet，我们也可以通过使用 spark.sql.sources.default 这个属性来设置默认的数据源。

```
val usersDF = spark.read.load("examples/src/main/resources/users.parquet")
usersDF.select("name",
"favorite_color").write.save("namesAndFavColors.parquet")
```

说明：
- spark.read.load 是加载数据的通用方法。
- df.write.save 是保存数据的通用方法。

1. 手动指定选项

可以手动给数据源指定一些额外的选项。数据源应该用全名称来指定，但是对一些内置的数据源也可以使用短名称，如 json、parquet、jdbc、orc、libsvm、csv、text 等。示例代码如下：

```
val peopleDF = spark.read.format("json").load("examples/src/main/resources/people.json")
peopleDF.select("name", "age").write.format("parquet").save("namesAndAges.parquet")
```

2. 在文件上直接运行 SQL

我们前面的做法都是使用 read API 先把文件加载到 DataFrame，然后再查询。其实也可以直接在文件上进行查询：

```
scala> spark.sql("select * from json.`examples/src/main/resources/
people.json`")
```

说明：json 表示文件的格式，后面的文件的具体路径需要用反引号（`）引起来。

3. 文件保存选项（SaveMode）

保存操作可以使用 SaveMode，用来指明如何处理数据，如表 4-1 所示。该操作使用 Data Frame 的 mode()方法来设置。

表4-1 文件保存选项

Scala/Java	Any Language	Meaning
SaveMode.ErrorIfExists(default)	"error"(default)	如果文件已经存在就抛出异常
SaveMode.Append	"append"	如果文件已经存在就追加
SaveMode.Overwrite	"overwrite"	如果文件已经存在就覆盖
SaveMode.Ignore	"ignore"	如果文件已经存在就忽略

注意：这些 SaveMode 都是没有加锁的，也不是原子操作。如果我们执行的是 Overwrite 操作，那么在写入新的数据之前会删除旧的数据。

4.3.2 加载 JSON 文件

Spark SQL 能够自动推测 JSON 数据集的结构，并将它加载为一个 Dataset[Row]。可以通过 SparkSession.read.json()去加载一个 JSON 文件，也可以通过 SparkSession.read.format("json").load()来加载。

注意：这个 JSON 文件不是一个传统的 JSON 文件，它每一行都得是一个完整的 JSON 串。

示例代码如下：

代码 4-3　DataSourceDemo.scala

```
{"name": "lisi", "age" : 20, "friends": ["lisi", "zs"]}
{"name": "zs", "age" : 30, "friends": ["lisi", "zs"]}
{"name": "wangwu", "age" : 15, "friends": ["lisi", "zs"]}

import org.apache.spark.sql.{DataFrame, Dataset, SparkSession}
object DataSourceDemo {
    def main(args: Array[String]): Unit = {
        val spark: SparkSession = SparkSession
            .builder()
            .master("local[*]")
            .appName("Test")
            .getOrCreate()
        import spark.implicits._
        val df: DataFrame = spark.read.json("target/classes/user.json")
        val ds: Dataset[User] = df.as[User]
        ds.foreach(user => println(user.friends(0)))
    }
```

```
    }
    case class User(name:String, age: Long, friends: Array[String])
```

4.3.3 读取 Parquet 文件

Parquet 是一种流行的列式存储格式，可以高效地存储具有嵌套字段的记录。Parquet 格式经常在 Hadoop 生态圈中被使用，它也支持 Spark SQL 的全部数据类型。Spark SQL 提供了直接读取和存储 Parquet 格式文件的方法。示例如下：

代码 4-4　DataSourceDemo2.scala

```
import org.apache.spark.sql.{DataFrame, Dataset, SaveMode, SparkSession}
object DataSourceDemo2 {
    def main(args: Array[String]): Unit = {
        val spark: SparkSession = SparkSession
            .builder()
            .master("local[*]")
            .appName("Test")
            .getOrCreate()
        import spark.implicits._
        val jsonDF: DataFrame = spark.read.json("target/classes/user.json")
        jsonDF.write.mode(SaveMode.Overwrite).parquet("target/classes/user.parquet")
        val parDF: DataFrame = spark.read.parquet("target/classes/user.parquet")
        val userDS: Dataset[User] = parDF.as[User]
        userDS.map(user => {user.name = "zl"; user.friends(0) = "志玲";user}).show()
    }
    case class User(var name:String, age: Long, friends: Array[String])
```

运行结果如图 4-8 所示。

```
Run:    DataSourceDemo
    2019-02-09 11:26:10,430 WARN [org.apache.parquet.hadoop.ParquetRecordRea
    +----+---+--------+
    |name|age| friends|
    +----+---+--------+
    |  zl| 20|[志玲, zs]|
    |  zl| 30|[志玲, zs]|
    |  zl| 15|[志玲, zs]|
    +----+---+--------+
```

图 4-8

注意：Parquet 格式的文件是 Spark 默认格式的数据源，所以当使用通用的方式时可以直接保存和读取，而不需要使用 format。spark.sql.sources.default 这个配置可以修改默认数据源。

4.3.4 JDBC

Spark SQL 也支持使用 JDBC 从其他的数据库中读取数据。使用 JDBC 数据源比使用 JdbcRDD

更优一些，这是因为它返回的结果直接就是一个 DataFrame，而 DataFrame 更加容易被处理或者与其他的数据源进行 join。

Spark SQL 可以通过 JDBC 从关系数据库中读取数据来创建 DataFrame，通过对 DataFrame 进行一系列的计算后，还可以将数据再写回关系数据库中。

注意：如果想在 spark-shell 操作 JDBC，那就需要把相关的 JDBC 驱动复制到 jars 目录下。

要从 JDBC 读取数据，首先需导入依赖：

```xml
<dependency>
    <groupId>mysql</groupId>
    <artifactId>mysql-connector-java</artifactId>
    <version>4.1.27</version>
</dependency>
```

1. 从 JDBC 读数据

从 JDBC 读数据可以使用通用的 load 方法，也可以使用 jdbc 方法。

（1）使用通用的 load 方法，示例代码如下：

代码 4-5　JDBCDemo.scala

```scala
import org.apache.spark.sql.SparkSession
object JDBCDemo {
  def main(args: Array[String]): Unit = {
    val spark: SparkSession = SparkSession
      .builder()
      .master("local[*]")
      .appName("Test")
      .getOrCreate()
    import spark.implicits._
    val jdbcDF = spark.read
      .format("jdbc")
      .option("url", "jdbc:mysql://hadoop201:3306/rdd")
      .option("user", "root")
      .option("password", "aaa")
      .option("dbtable", "user")
      .load()
    jdbcDF.show
  }
}
```

（2）使用 jdbc 方法，代码如下：

代码 4-6　JDBCDemo2.scala

```scala
import java.util.Properties
import org.apache.spark.sql.{DataFrame, SparkSession}
object JDBCDemo2 {
  def main(args: Array[String]): Unit = {
    val spark: SparkSession = SparkSession
```

```
            .builder()
            .master("local[*]")
            .appName("Test")
            .getOrCreate()
        val props: Properties = new Properties()
        props.setProperty("user", "root")
        props.setProperty("password", "aaa")
        val df: DataFrame = spark.read.jdbc("jdbc:mysql://hadoop201: 3306/rdd", "user", props)
        df.show
    }
}
```

2. 向 jdbc 写入数据

向 jdbc 写入数据也分两种方法：通用 write.save 和 write.jdbc。示例代码如下：

代码 4-7　JDBCDemo3.scala

```
import java.util.Properties
import org.apache.spark.rdd.RDD
import org.apache.spark.sql.{DataFrame, Dataset, SaveMode, SparkSession}
object JDBCDemo3 {
    def main(args: Array[String]): Unit = {
        val spark: SparkSession = SparkSession
            .builder()
            .master("local[*]")
            .appName("Test")
            .getOrCreate()
        import spark.implicits._
        val rdd: RDD[User1] = spark.sparkContext.parallelize(Array(User1("lisi", 20), User1("zs", 30)))
        val ds: Dataset[User1] = rdd.toDS
        ds.write
            .format("jdbc")
            .option("url", "jdbc:mysql://hadoop201:3306/rdd")
            .option("user", "root")
            .option("password", "aaa")
            .option("dbtable", "user")
            .mode(SaveMode.Append)
            .save()
        val props: Properties = new Properties()
        props.setProperty("user", "root")
        props.setProperty("password", "aaa")
        ds.write.mode(SaveMode.Append).jdbc("jdbc:mysql://hadoop201:3306/rdd", "user", props)
    }
}
case class User1(name: String, age: Long)
```

第 5 章

Kafka 实战

Kafka 是一个分区的、多副本的、多订阅者的、基于 ZooKeeper 协调的分布式日志系统（也可以当作消息队列（Message Queuing）系统，如图 5-1 所示。Kafka 常用于 Web/Nginx 日志、访问日志、消息服务等，Linkedin 于 2010 年将它贡献给了 Apache 基金会并成为顶级开源项目。

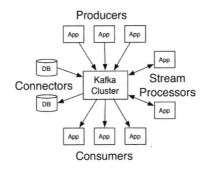

图 5-1

Kafka 主要应用场景：日志收集系统和消息系统。

Kafka 主要设计目标如下：

- 以时间复杂度为 O(1) 的方式提供消息持久化能力，即使对 TB 级以上数据也能保证常数时间的访问性能。
- 高吞吐率。即使在非常廉价的商用机器上也能做到单机支持每秒 100K 条消息的传输。
- 支持 Kafka Server 间的消息分区及分布式消费，同时保证每个分区内的消息顺序传输。
- 同时支持离线数据处理和实时数据处理。
- 支持在线水平扩展。

本章主要知识点：

- Kafka 的特点
- Kafka 术语
- Kafka 单机部署
- Kafka 集群部署

下载的 Kafka 版本需要与我们当前安装的 Scala 版本对应。本书选择 kafka_2.12-2.7.0.tgz，其中 2.12 是 Scala 的版本，2.7 是 Kafka 的版本。下载地址为：http://kafka.apache.org。

5.1 Kafka 的特点

Kafka 的特点如下：

1. 解耦

消息系统在处理过程中插入了一个隐含的、基于数据的接口层，两边的处理过程都要实现这一接口。这允许我们独立地扩展或修改两边的处理过程，只要确保它们遵守同样的接口约束。

2. 冗余（副本）

有些情况下，处理数据的过程会失败，除非数据被持久化，否则将造成数据丢失。消息队列把数据进行持久化直到它们被完全处理，通过这一方式规避了数据丢失的风险。许多消息队列所采用的"插入-获取-删除"范式中，在把一个消息从队列中删除之前，需要处理系统明确指出该消息已经被处理完毕，从而确保数据被安全地保存直到使用完毕。

3. 扩展性

因为消息队列解耦了处理过程，所以增大消息入队和处理的频率变得很容易，只要另外增加处理过程即可。不需要改变代码、不需要调节参数，扩展就像调电脑播放器按钮一样简单。

4. 灵活性&峰值处理能力

在访问量急剧增加的情况下，应用仍然需要继续发挥作用，但是这样的突发流量并不常见。如果以能处理这类峰值访问为标准来投入资源随时待命，无疑是巨大的浪费。使用消息队列能够使关键组件顶住突发的访问压力，而不会因为突发的超负荷的请求而完全崩溃。

5. 可恢复性

系统的一部分组件失效不会影响到整个系统。消息队列降低了进程间的耦合度，所以即使一个处理消息的进程挂掉，加入队列中的消息仍然可以在系统恢复后被处理。

6. 顺序保证

在大多使用场景中，数据处理的顺序都很重要。大部分消息队列本来就是排序的，并且能保证数据会按照特定的顺序来处理。Kafka 保证了一个分区内消息的有序性。

7. 缓冲

在任何重要的系统中，都会有需要不同的处理时间的元素，例如加载一幅图片比应用过滤器花费更少的时间。消息队列通过一个缓冲层来帮助任务最高效率地执行——写入队列的处理会尽可能地快速。该缓冲有助于控制和优化数据流经过系统的速度。

8. 异步通信

很多时候，用户不想也不需要立即处理消息。消息队列提供了异步处理机制，允许用户把一个消息放入队列，但并不立即处理它。想向队列中放入多少消息就放多少，然后在需要的时候再去处理它们。

5.2 Kafka 术语

本节主要介绍 Kafka 中的常用术语。

1. broker

Kafka 集群包含一个或多个服务器，服务器节点称为 broker。broker 存储 topic 的数据。

如果某 topic 有 N 个 partition，集群有 N 个 broker，那么每个 broker 存储该 topic 的一个 partition。

如果某 topic 有 N 个 partition，集群有 $(N+M)$ 个 broker，那么其中有 N 个 broker 存储该 topic 的一个 partition，剩下的 M 个 broker 不存储该 topic 的 partition 数据。

如果某 topic 有 N 个 partition，集群中 broker 数目少于 N 个，那么一个 broker 存储该 topic 的一个或多个 partition。在实际生产环境中，应尽量避免这种情况的发生，这种情况容易导致 Kafka 集群数据不均衡。

2. topic

每条发布到 Kafka 集群的消息都有一个类别，这个类别被称为 topic，类似于数据库的表名。物理上不同 topic 的消息被分开存储，逻辑上一个 topic 的消息虽然保存于一个或多个 broker 上，但用户只需指定消息的 topic 即可生产或消费数据而不必关心数据存于何处。

3. partition

topic 中的数据分割为一个或多个 partition，也称之为分区。topic 是逻辑概念，partition 是物理概念，生产者只需要关心数据发给哪一个 topic，消费者也只关心自己读取了哪一个 topic。每个 topic 至少有一个 partition。每个 partition 中的数据使用多个 segment 文件存储。partition 中的数据是有序的，不同 partition 间的数据丢失了数据的顺序，如果 topic 有多个 partition，那么消费数据时就不能保证数据的顺序。在需要严格保证消息的消费顺序的场景下，需要将 partition 数目设为 1。

4. Producer

Producer（生产者），即数据的发布者，该角色将消息发布到 Kafka 的 topic 中。broker 接收到生产者发送的消息后，将该消息追加到当前用于追加数据的 segment 文件中。生产者发送的消息存储到一个 partition 中，生产者也可以指定数据存储的 partition。

5. Consumer

Consumer（消费者）可以从 broker 中读取数据。消费者可以消费多个 topic 中的数据。

6. Consumer Group

每个 Consumer 属于一个特定的 Consumer Group，即可为每个 Consumer 指定 group name（组名）；若不指定 group name，则属于默认的 group。

7. Leader

每个 partition 有多个副本，其中有且仅有一个作为 Leader（领导者），Leader 是当前负责数据读写的 partition。

8. Follower

Follower（跟随者）跟随 Leader，所有写请求都通过 Leader 路由，数据变更会广播给所有 Follower，Follower 与 Leader 保持数据同步。如果 Leader 失效，则从 Follower 中选举出一个新的 Leader。当 Follower 与 Leader 挂掉、卡住或者同步太慢时，Leader 会把这个 Follower 从 "in sync replicas（ISR）" 列表中删除，重新创建一个 Follower。

5.3 Kafka 单机部署

Kafka 可以在一台机器上安装多个，只要在启动时，指定不同的 server.properties 文件就可以了，此文件中配置的是主机名和端口。

单机（也称单节点）部署就是在单个主机上安装 Kafka，同时需要 ZooKeeper 的支持。

Kafka 单机部署的操作步骤如下：

步骤01 安装与配置 ZooKeeper。

在安装 ZooKeeper 之前，确认已经安装了 JDK，并正确配置了 JAVA_HOME 和 PATH 环境变量。单节点安装只要解压 ZooKeeper 并配置 zoo.cfg 文件，然后修改 dataDir 数据保存目录。

（1）下载 ZooKeeper。可以通过 wget 下载 ZooKeeper。如果没有安装 wget 可以通过 yum install -y wget 安装此软件，也可以通过 xftp 上传已经下载好的 ZooKeeper 压缩包文件。

```
$ https://www.apache.org/dyn/closer.lua/zookeeper/zookeeper-3.8.1/apache-zookeeper-3.8.1-bin.tar.gz
```

（2）将 ZooKeeper 压缩包文件解压到指定的目录下或当前目录下，并把目录名简化一下：

```
$ tar -zxvf ~/apache-zookeeper-3.8.1-bin.tar.gz -C /app/
$ mv apache-zookeeper-3.8.1-bin/ zookeeper-3.8.1
```

（3）复制示例配置文件。进入 ZooKeeper 目录，将 zoo_sample.cfg 复制为 zoo.cfg 文件。

```
$ cp zoo_sample.cfg zoo.cfg
```

zoo.cfg 配置如下：

```
tickTime=2000
initLimit=10
```

```
syncLimit=5
dataDir=/app/datas/zookeeper
clientPort=2181
# ZooKeeper 与 Spark 端口相同，需要修改，此值默认为 8080
admin.serverPort=9999
```

（4）启动 ZooKeeper。ZooKeeper 默认使用端口 2181。

启动 zk server：

```
[hadoop@server201 /app]$/app/zookeeper-3.8.1/bin/zkServer.sh start
ZooKeeper JMX enabled by default
Using config: /app/zookeeper-3.8.1/bin/../conf/zoo.cfg
Mode: standalone
```

查看 ZooKeeper 的进程，其中 QuorumPeerMain 为 zkServer 的进程。

```
$ jps
2490 QuorumPeerMain
```

（5）登录客户端。zkCli.sh 在 ZOOKEEPER_HOME/bin 目录下，可以在脚本中简单地配置一下 ZooKeeper 的环境变量：

```
export ZOOKEEPER_HOME=/app/zookeeper-3.8.1
export PATH=.:$PATH:$ZOOKEEPER_HOME/bin
```

配置好环境变量之后，可以在任意的目录下执行 zkCli.sh 命令以登录 zkServer。

步骤 02 安装 Kafka。

（1）解压 Kafka 安装文件：

```
$ tar -zxvf kafka-2.12.0_2.7.0.tgz -C /app/
```

（2）配置 server.properties 文件，因为 ZooKeeper 也是单机安装，所以它只有一个节点：

```
zookeeper.connect=hadoop201:2181
```

（3）配置环境变量：

```
$ vim /etc/profile
export KAFKA_HOME=/app/kafka-2.12.0_2.7.0
export PATH=$PATH:$KAFKA_HOME/bin
```

（4）让环境变量生效：

```
$source  /etc/profile
```

（5）查看 Kafka 的版本：

```
[hadoop@server201 bin]$ kafka-server-start.sh --version
[2021-03-26 18:56:35,340] INFO Registered kafka:type=kafka.Log4jController
MBean (kafka.utils.Log4jControllerRegistration$)
2.7.0 (Commit:448719dc99a19793)
```

步骤 03 启动 Kafka。

（1）使用 kafka-server-start.sh 启动 Kafka 进程：

```
$ kafka-server-start.sh /app/kafka-2.0/config/server.properties &
```

或使用 --daemon 参数启动（建议）：

```
$ bin/kafka-server-start.sh -daemon config/server.properties
```

（2）查看进程：

```
[hadoop@server201 app]$ jps
1381 QuorumPeerMain
1894 Kafka
```

（3）查看 ZooKeeper 中已经存在的节点：

```
[zk: localhost:2181(CONNECTED) 0] ls /
[cluster, controller_epoch, controller, brokers, zookeeper, admin,
isr_change_notification, consumers, log_dir_event_notification,
latest_producer_id_block, config]
[zk: localhost:2181(CONNECTED) 1] ls /brokers
[ids, topics, seqid]
[zk: localhost:2181(CONNECTED) 2] ls /brokers/topics
[]
```

可以看到，cluster、brokers 已经注册到 ZooKeeper 里面了。

步骤 04 注册一个 topic。

（1）Kafka-topics.sh 可以对 topic 进行创建、删除等。通过 —help 选项可以查看此命令的帮助。

```
[hadoop@server201 bin]$ kafka-topics.sh --help
This tool helps to create, deslete, describe, or change a topic.
Option                                  Description
------                                  -----------
--alter                                 Alter the number of partitions,
                                        replica assignment, and/or
...
```

（2）注册或创建一个 topic：

```
$ kafka-topics.sh --create --bootstrap-server server201:9092--topic mytopic \
> --replication-factor 1 --partitions 1
Created topic "mytopic".
```

（3）查看这个 topic：

```
$ kafka-topics.sh --bootstrap-server server201:9092 --describe
Topic:mytopic   PartitionCount:1    ReplicationFactor:1 Configs:
Topic: mytopic  Partition: 0    Leader: 0   Replicas: 0 Isr: 0
```

（4）使用 list 列出所有 topics：

```
[hadoop@server201 bin]$ kafka-topics.sh --zookeeper server201:2181 --list
mytopic
```

topic 在 Kafka 中是主题的意思，生产者将消息发送到主题，消费者再订阅相关的主题，并从主题上拉取消息。在创建 topic 的时候，有两个参数是需要填写的，那就是 partitions 和 replication-factor。

- partitions

partitions 是主题分区数，在创建 topic 时，通过 --partitions 参数来指定。

Kafka 通过分区策略将不同的分区分配在一个集群中的 broker 上，一般会分散在不同的 broker 上，当只有一个 broker 时，所有的分区就只分配到该 broker 上。消息会通过负载均衡发布到不同的分区上，消费者通过监测偏移量来获取哪个分区有新数据，从而从该分区上拉取消息数据。

分区数越多，在一定程度上会提升消息处理的吞吐量，因为 Kafka 是基于文件进行读写的，因此需要打开更多的文件句柄，也会增加一定的性能开销。如果分区过多，那么日志分段也会很多，写的时候由于是批量写，因此就会变成随机写了，随机 I/O 这个操作对性能影响很大，所以一般来说 Kafka 不能有太多的 partition。

图 5-2 演示了当 partition 与 broker 相同时的分配情况，设置 topic-1 的 partitions 个数为 2，broker 的个数也为 2，采用均匀分配策略。当 broker 和 partitions 配置数一样时，partition 就均匀分布在不同的 broker 上。

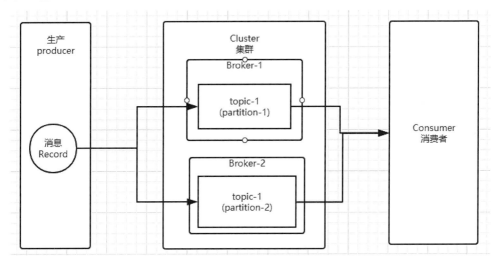

图 5-2

图 5-3 演示了 partition 与 broker 不相同时的分配情况（建议不要出现这种情况），topic 的 partition 个数为 3，但 broker 的个数为 2。

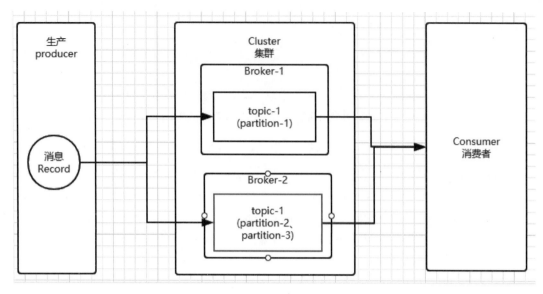

图 5-3

- replication-factor

replication-factor 用来设置主题的副本数。每个主题可以有多个副本，副本位于集群中不同的 broker 上，也就是说副本的数量不能超过 broker 的数量，否则创建主题时会失败。

假如我们目前已经有 3 个 broker（即集群节点数量为 3），则创建以下 toppic：

```
$kafka-topics.sh --create --topic topicA \
--boostrap-server server202,server203,server201:9092 \
--replication-factor 2 --partitions 3
Created topic topicA.
```

上面的脚本创建了一个 topicA，指定 topicA 的每一个分区副本数量为 2，查看这个 topicA 的信息为：

```
$ kafka-topics.sh --describe \
--boostrap-server server202,server203,server201:9092 --topic topicA \
topicA   PartitionCounter:3      ReplicationFactor:2   Configs:
 Topic:topicA  Partition:0     Leader:201     Replicas:201,203    Isr:201,203
 Topic:topicA  Partition:0     Leader:202     Replicas:201,202    Isr:201,202
 Topic:topicA  Partition:0     Leader:203     Replicas:202,203    Isr:203,202
```

可一看到，每一个分区（如 partition:0）在 203 和 201 服务器上保存了副本（一共两个副本）。

再来测试一下：

如果在集群为 3 的 broker 上指定 partitions=1 和 replication-factor=1：

```
    [hadoop@server101 ~]$ kafka-topics.sh --create --topic topicB
--boostrap-server server101,server102,server103:9092 --replication-factor 1
--partitions 1
    Created topic topicB.
```

则只会在一个 broker 上创建 partition 且只会有一个副本，这样做很不合理。

查看:

```
[hadoop@server101 ~]$ kafka-topics.sh --describe \
--boostrap-server server101,server102,server103:9092 --topic topicB
Topic:topicB    PartitionCount:1    ReplicationFactor:1 Configs:
Topic: topicB   Partition: 0    Leader: 101 Replicas: 101   Isr: 101
```

可见,虽然分区都在集群上,但都保存到一个 borker 中。

最后我们给出配置建议:

- replication-factor 应该小于或等于 broker。
- partitions 等于 broker 的数量。

步骤 05 创建发布者和消费者。

(1) 创建发布者并发布一些数据:

```
[hadoop@server201 app]$ kafka-console-producer.sh \
--broker-list server201:9092 --topic mytopic
>Jack Mary
>rose
>Jack
```

(2) 创建消费者,用于读取发布者发布的数据并显示到控制台:

```
[hadoop@server201 app]$ kafka-console-consumer.sh \
--bootstrap-server server201:9092 \
> --topic mytopic --from-beginning
Jack Mary
rose
Jack
```

步骤 06 使用 Java 代码访问 topic。

(1) 请确定之前名称为 mytopic 的 topic 依然存在,可以通过 --list 查看:

```
[hadoop@server201 bin]$ kafka-topics.sh --boostrap-server server201:2181
--list
mytopic
```

(2) 在 pom.xml 添加 Kakfa 的依赖:

```xml
<dependency>
    <groupId>org.apache.kafka</groupId>
    <artifactId>kafka_2.12</artifactId>
    <version>2.7.0</version>
</dependency>
```

(3) 创建发布者并启动如下代码:

代码 5-1 MyProducer.java

```
package org.hadoop.kakfa;
```

```java
import org.apache.kafka.clients.producer.KafkaProducer;
import org.apache.kafka.clients.producer.Producer;
import org.apache.kafka.clients.producer.ProducerConfig;
import org.apache.kafka.clients.producer.ProducerRecord;
import org.apache.kafka.common.serialization.StringSerializer;
import java.util.Properties;
import java.util.Scanner;
public class MyProducer {
    public static void main(String[] args) throws Exception {
        Properties properties = new Properties();
        properties.put("bootstrap.servers", "server201:9092");
        //以下是可选属性
        properties.put("acks", "all");
        properties.put("retries", 0);
        properties.put("batch.size", 16384);
        properties.put("linger.ms", 1);
        properties.put("buffer.memory", 33554432);
        //key.serializer=org.apache.kafka.common.serialization.StringSerializer
        properties.put(ProducerConfig.KEY_SERIALIZER_CLASS_CONFIG, StringSerializer.class.getName());
        //value.serializer=org.apache.kafka.common.serialization.StringSerializer
        properties.put(ProducerConfig.VALUE_SERIALIZER_CLASS_CONFIG, StringSerializer.class.getName());
        Producer<String, String> producer = new KafkaProducer<>(properties);
        Scanner sc = new Scanner(System.in);//接收用户的输入
        for (int i = 0; ; i++) {
            String line = sc.nextLine();
            if (line.equals("exit")) {
                break;
            }
            ProducerRecord<String, String> record =
                new ProducerRecord<>("mytopic", "key", line);
            //发送数据
            producer.send(record);
        }
        producer.close();
    }
}
```

（4）启动后，我们输入一些数据：

```
This is my First Line
This is My Second Line
```

（5）创建消费接收生产者发送的数据，然后查看接收到的数据。代码如下：

代码 5-2 MyConsumer.java

```
package org.hadoop.kakfa;
import org.apache.kafka.clients.consumer.*;
```

```java
import org.apache.kafka.common.serialization.StringDeserializer;
import java.time.Duration;
import java.util.Arrays;
import java.util.Properties;
public class MyConsumer {
    public static void main(String[] args) {
        Properties properties = new Properties();
        properties.put("bootstrap.servers", "server201:9092");
        //以下是可选属性
        properties.put("group.id", "test");
        properties.put("enable.auto.commit", "true");
        properties.put("auto.commit.interval.ms", "1000");
        properties.put("session.timeout.ms", "30000");
        //key.serializer=org.apache.kafka.common.serialization.StringSerializer
        //配置如何解析 key-value
        properties.put(ConsumerConfig.KEY_DESERIALIZER_CLASS_CONFIG, StringDeserializer.class.getName());
        properties.put("value.deserializer", "org.apache.kafka.common.serialization.StringDeserializer");
        Consumer<String, String> consumer = new KafkaConsumer<String, String>(properties);
        //监听的 topic 名称
        consumer.subscribe(Arrays.asList("mytopic"));
        while (true) {
            ConsumerRecords<String, String> records = consumer.poll(Duration.ofSeconds(2));
            for (ConsumerRecord<String, String> record : records) {
                System.err.println(">>>>>>>>>>:" + record.key() + "\t" + record.value());
            }
        }
    }
}
```

启动后，接收到的数据如下：

```
>>>>>>>>>>:key  This is my First Line
>>>>>>>>>>:key  This is My Second Line
```

5.4　Kafka 集群部署

Kafka 集群部署的核心是配置 config/server.properties 文件，代码如下：

```
broker.id=101   #唯一地标识 int 类型
listeners=PLAINTEXT://server101:9092  #指定本机的地址
log.dirs=/home/isoft/logs/kafka   #指定一个空的已经存在的目录
#指定外部 ZooKeeper 的地址，可选地指定一个子目录只保存 Kafka 的信息
zookeeper.connection=server101,server102,server103:2181/kafka
```

Kafka 集群部署规划如表 5-1 所示。

表5-1　Kafka集群部署规划

主机/IP/虚拟机名	软　件	进　程	标　识
server101 192.168.56.101 CentOS7-101	zookeeper-3.5.5 kafka_2.11_2.3.1 jdk-1.8	QuorumPeerMan Kafka	myid=101 broker.id=101
server102 192.168.56.102 CentOS7-102	同上	QuorumPeerMan Kafka	myid=102 broker.id=102
server103 192.168.56.103 CentOS7-103	同上	QuorumPeerMan Kafka	myid=103 broker.id=103

操作步骤如下：

步骤01 配置 ZooKeeper 集群并启动。

（1）解压 ZooKeeper：

```
$ tar -zxvf ~/apache-zookeeper-3.5.5-bin.tar.gz -C /app/
```

（2）修改 ZooKeeper 的配置文件<zookeeper_home>/conf/zoo.cfg：

```
tickTime=2000
initLimit=10
syncLimit=5
dataDir=/home/isoft/datas/zk
clientPort=2181
admin.serverPort=9999
server.101=server101:2888:3888
server.102=server102:2888:3888
server.103=server103:2888:3888
```

（3）分别在每台服务器的 home/isoft/datas/zk 目录下添加 myid，内容为当前 ZooKeeper 的 id：

```
[hadoop@server101 ~]$ echo 101 > /home/isoft/datas/zk/myid
[hadoop@server102 ~]$ echo 102 > /home/isoft/datas/zk/myid
[hadoop@server103 ~]$ echo 103 > /home/isoft/datas/zk/myid
```

（4）可以使用脚本一次启动所有 ZooKeeper 集群：

```
#!/bin/bash
if [ $# -lt 1 ]; then
    echo "用法: $0 start | stop | status"
    exit 1
fi
hosts=(server101 server102 server103)
cmd=$1
for host in ${hosts[@]};do
    script="ssh ${host} zkServer.sh ${cmd}"
    echo $script
```

```
    eval $script
  done
exit 0
```

步骤 02 安装 Kafka。

(1) 解压 Kafka：

```
$ tar -zxvf /home/hadoop/kafka_2.12.0-2.7.0.tgz -C /app/
```

(2) 修改目录名称：

```
$ mv kafka_2.20.0-2.7.0  kafka-2.7.0
```

(3) 修改 Kafka 配置文件 server.properties：

```
broker.id=101   #根据主机的不同，分别设置 broker.id=102,broker.id=103
#根据主机不同，分别设置 server102:9092,server103:9092
listeners=PLAINTEXT://server101:9092
log.dirs=/home/hadoop/logs/kafka
#可以声明一个子目录，便于管理
zookeeper.connect=server101,server102,server103:2181/kafka
```

(4) 将 Kafka 分发到其他主机的相同目录下并按上一步的说明做修改：

```
$ scp -r kafka-2.7.0 server102:/app/
$ scp -r kafka-2.7.0 server103:/app/
```

(5) 配置所有主机的环境变量：

```
export KAFKA_HOME=/app/kafka-2.7.0
export PATH=$PATH:$KAFKA_HOME/bin
```

步骤 03 启动 Kafka。

(1) 逐台主机上启动 Kafka 服务器：

```
$ kafka-server-start.sh  /app/kafka-2.7.0/config/server.properties -daemon
```

(2) 使用脚本一次性启动：

```
#!/bin/bash
if [ $# -lt 1 ]; then
    echo "使用方法: $0 start | stop"
    exit 1
fi
cmd=$1
servers=(server101 server102 server103)
if [ $cmd == 'start' ]; then
    echo "启动"
    for host in ${servers[@]};
    do
        script="ssh $host kafka-server-start.sh -daemon /app/kafka-2.7.0/config/server.properties"
        echo $script
        eval $script
```

```
        done
        exit 0
    elif [ $cmd == 'stop' ]; then
        echo "停止"
        for host in ${servers[@]};do
            script="ssh $host kafka-server-stop.sh
/app/kafka-2.7.0/config/server.properties"
            echo $script
            eval $script
        done
        exit 0
    else
        echo "错误的参数"
        exit 1
    fi
```

（3）Kafka 服务器都启动以后，查看一下所有 broker 的信息。版本信息如下：

```
[hadoop@server101 ~]$ kafka-broker-api-versions.sh --version
2.7.0 (Commit:18a913733fb71c01)
```

输入所有的服务器地址和输入一个服务器地址返回的信息都是一样的：

```
$ kafka-broker-api-versions.sh \
--bootstrap-server server101:9092,server102:9092,server103:9092
```

以下是输入一个服务器地址：

```
[hadoop@server101 ~]$ kafka-broker-api-versions.sh --bootstrap-server
server101:9092
```

返回的信息如下：

```
server103:9092 (id: 103 rack: null) -> (
    Produce(0): 0 to 7 [usable: 7],...
server101:9092 (id: 101 rack: null) -> (
    Produce(0): 0 to 7 [usable: 7],...
server102:9092 (id: 102 rack: null) -> (
    Produce(0): 0 to 7 [usable: 7],...
```

步骤04 创建并查看 topic。

（1）创建 topic：

```
$ kafka-topics.sh --create --topic two \
--bootstrap-server server101:9092,server102:9092,server103:9092 \
--partitions 3 --replication-factor 3
```

（2）查看 topic：

```
$ kafka-topics.sh --describe --topic two --bootstrap-server server101:9092
Topic:two    PartitionCount:3    ReplicationFactor:3
Configs:segment.bytes=1073741824
    Topic: two    Partition: 0    Leader: 101 Replicas: 101,103,102    Isr:
101,103,102
```

```
        Topic: two  Partition: 1    Leader: 103 Replicas: 103,102,101   Isr: 103,102,101
        Topic: two  Partition: 2    Leader: 102 Replicas: 102,101,103   Isr: 102,101,103
```

步骤 05 发布订阅。

(1) 创建发布者:

```
$ kafka-console-producer.sh --topic two \
--broker-list server101:9092,server102:9092,server103:9092
>Jack
>mary
>Rose
```

(2) 创建消费者:

```
$ kafka-console-consumer.sh --topic two \
--bootstrap-server server101:9092,server102:9092,server103:9092 --from-beginning
   Rose
   mary
   Jack
```

第 6 章

Spark Streaming 实时计算

近年来，在 Web 应用、网络监控、传感监测、电信金融、生产制造等领域增强了对数据实时处理的需求，而 Spark 中的 Spark Streaming 实时计算框架就是为了满足数据实时处理的需求而设计的。在电子商务领域中，淘宝、京东网站从用户点击的行为和浏览的历史记录中发现用户的购买意图和兴趣，然后通过 Spark Streaming 实时计算框架进行分析处理，为用户推荐相关商品，从而有效地提高商品的销售量，同时也增加了用户的满意度，可谓是"一举两得"。

本章主要知识点：

- Spark Streaming 概述
- DStream 入门
- DStream 创建
- DStream 实战
- Structured Streaming 应用

6.1　Spark Streaming 概述

本节主要介绍 Spark Streaming 相关内容，包括 Spark Streaming 的概念、特点和架构。

6.1.1　Spark Streaming 是什么

Spark Streaming 是 Spark 核心 API 的扩展，是用于构建弹性、高吞吐量、容错的在线数据流的流式处理程序。一句话概括就是 Spark Streaming 用于流式数据的处理。数据可以来源于多种数据源：Kafka、Flume、Kinesis 或者 TCP 套接字。接收到的数据可以使用 Spark 的原语来处理，尤其是那些高阶函数，例如 map、reduce、join 和 window。最终，被处理的数据可以发布到 HDFS、数据库或者在线 Dashboards 上，如图 6-1 所示。

图 6-1

另外，Spark Streaming 也能和机器学习以及 Graphx 完美融合。

在 Spark Streaming 中，处理数据的单位是一批而不是单条，而数据采集却是逐条进行的，因此 Spark Streaming 系统需要设置间隔，使得数据汇总到一定量后再一并操作，这个间隔就是批处理间隔。批处理间隔是 Spark Streaming 的核心概念和关键参数，它决定了 Spark Streaming 提交作业的频率和数据处理的延迟，同时也影响着数据处理的吞吐量和性能。Spark Streaming 处理数据的流程如图 6-2 所示。

图 6-2

Spark Streaming 提供了一个高级抽象——Discretized Stream（DStream），DStream 表示一个连续的数据流。

DStream 可以由来自数据源的输入数据流来创建，也可以通过在其他的 DStream 上应用一些高阶操作来得到。在 Spark Streaming 框架内部，一个 DSteam 是由一个个 RDD 序列来表示的。

6.1.2 Spark Streaming 特点

Spark Streaming 具有以下 3 个优点。

（1）易用，可通过高阶函数来构建应用，如图 6-3 所示。

图 6-3

（2）容错，如图 6-4 所示。

图 6-4

（3）容易整合到 Spark 体系中，如图 6-5 所示。

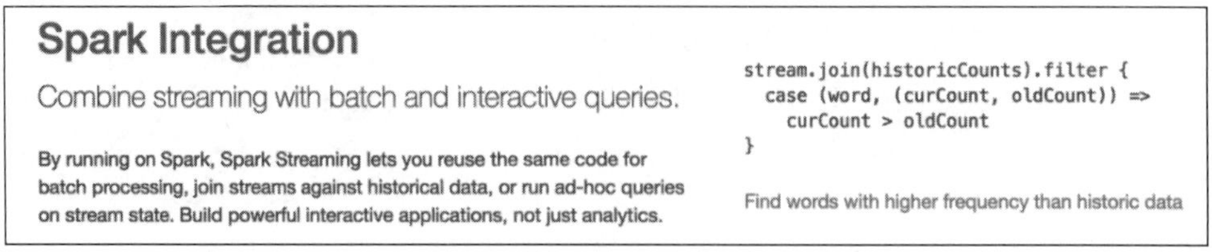

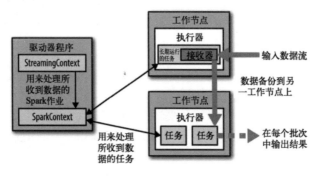

图 6-5

当然，Spark Streaming 也有缺点，Spark Streaming 是一种"微量批处理"架构，和其他基于"一次处理一条记录"架构的系统相比，它的延迟会相对高一些。

6.1.3　Spark Streaming 架构

Spark Streaming 架构如图 6-6 所示。

图 6-6

Spark 1.5 以前的版本，用户如果要限制 Receiver 的数据接收速率，可以通过设置静态配制参数 spark.streaming.receiver.maxRate 的值来实现。此举虽然可以通过限制接收速率来适配当前的处理能力，防止内存溢出，但也会引入其他问题，比如当 Producer 数据生产高于 maxRate，当前集群处理能力也高于 maxRate 时，就会造成资源利用率下降等问题。

为了更好地协调数据接收速率与资源处理能力，从 Spark 1.5 之后的版本开始，Spark Streaming 可以动态控制数据接收速率来适配集群数据处理能力——背压机制（Spark Streaming Backpressure），根据 JobScheduler 反馈作业的执行信息来动态调整 Receiver 数据接收率。通过属性 spark.streaming.backpressure.enabled 来控制是否启用背压机制，默认值为 false，即不启用。

6.2　DStream 入门

本节通过介绍最基本的 DStream 案例——WordCount，让读者快速进入 DStream。

6.2.1 WordCount 案例

1. 需求

使用 Netcat 工具向 9999 端口不断地发送数据，通过 Spark Streaming 读取端口数据并统计不同单词出现的次数。

2. 添加依赖

代码如下：

```xml
<dependency>
    <groupId>org.apache.spark</groupId>
    <artifactId>spark-streaming_2.11</artifactId>
    <version>2.1.1</version>
</dependency>
```

3. 编写代码

代码如下：

代码 6-1　StreamingWordCount.scala

```scala
import org.apache.spark.streaming.dstream.{DStream, ReceiverInputDStream}
import org.apache.spark.streaming.{Seconds, StreamingContext}
import org.apache.spark.{SparkConf, SparkContext}
object StreamingWordCount {
    def main(args: Array[String]): Unit = {
        val conf = new SparkConf().setAppName("StreamingWordCount").setMaster("local[*]")
        //1. 创建 Spark Streaming 的入口对象——StreamingContext。参数 2 表示事件间隔。其中 StreamingWordCount 对象内部会创建 SparkContext
        val ssc = new StreamingContext(conf, Seconds(3))
        // 2. 创建一个 DStream
        val lines: ReceiverInputDStream[String] = ssc.socketTextStream("hadoop201", 9999)
        // 3. 一个一个的单词
        val words: DStream[String] = lines.flatMap(_.split("""\s+"""))
        // 4. 单词形成元组
        val wordAndOne: DStream[(String, Int)] = words.map((_, 1))
        // 5. 统计单词的个数
        val count: DStream[(String, Int)] = wordAndOne.reduceByKey(_ + _)
        //6. 显示
        println("aaa")
        count.print
        //7. 开始接收数据并计算
        ssc.start()
        //8. 等待计算结束(要么手动退出，要么出现异常)才退出主程序
        ssc.awaitTermination()
    }
}
```

4. 测试

（1）在 hadoop201 系统上启动 Netcat：

```
nc -lk 9999
```

（2）可以打包到 Linux 启动 WordCount，也可以在 IDEA 上直接启动。

（3）查看输出结果，如图 6-7 所示，每 3 秒统计一次数据的输入情况。

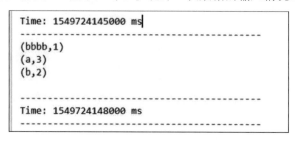

图 6-7

注意：如果日志太多，那么可以把日志级别修改为 ERROR。

5. 注意事项

（1）StreamingContext 一旦启动，则不能再添加新的 streaming computations。

（2）StreamingContext 一旦停止（StreamingContext.stop()），则再也不能重启。

（3）在一个 JVM 内，同一时间只能启动一个 StreamingContext。

（4）使用 stop() 的方式停止 StreamingContext 的同时，也会把 SparkContext 停掉。如果仅仅想停止 StreamingContext，则应该使用 stop(false)。

（5）一个 SparkContext 可以重用去创建多个 StreamingContext，前提是以前的 StreamingContext 已经停掉，并且 SparkContext 没有被停掉。

6.2.2　WordCount 案例解析

DStream 是 Spark Streaming 提供的基本抽象，表示持续性的数据流，可以来自输入数据，也可以由其他 DStream 转换得到。在框架内部，一个 DSteam 用连续的一系列的 RDD 来表示，如图 6-8 所示，在 DStream 中的每个 RDD 包含一个确定时间段的数据。

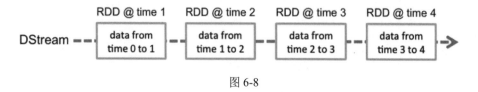

图 6-8

对 DStream 的任何操作都会转换成对它里面的 RDD 的操作，比如 6.2.1 节的 WordCount 案例，flatMap 是应用在 lines DStream 的每个 RDD 上，然后生成了 words DStream 中的 RDD，流程如图 6-9 所示。

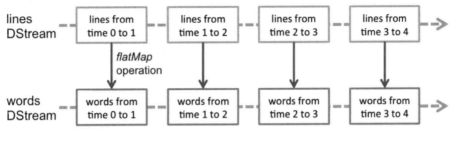

图 6-9

对这些 RDD 的转换是交给 Spark 引擎来计算的，原理如图 6-10 所示。DStream 的操作隐藏了大多数的细节，然后给开发者提供了方便使用的高级 API。

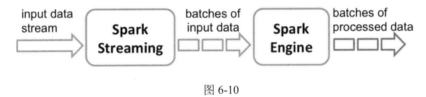

图 6-10

6.3　DStream 创建

Spark Streaming 原生支持一些不同的数据源，本节主要介绍如何使用不同数据源创建 DStream。

6.3.1　RDD 队列

1. 用法及说明

测试过程中，可以通过使用 ssc.queueStream(queueOfRDDs)来创建 DStream，每一个推送到这个队列中的 RDD 都会作为一个 DStream 被处理。

2. 需求

循环创建几个 RDD，将 RDD 放入队列。通过 Spark Streaming 创建 Dstream，计算 WordCount。

3. 代码

代码 6-2　RDDQueueDemo.scala

```scala
import org.apache.spark.SparkConf
import org.apache.spark.rdd.RDD
import org.apache.spark.streaming.dstream.InputDStream
import org.apache.spark.streaming.{Seconds, StreamingContext}
import scala.collection.mutable
/**  * mrchi  */
object RDDQueueDemo {
    def main(args: Array[String]): Unit = {
        val conf = new
```

```
SparkConf().setAppName("RDDQueueDemo").setMaster("local[*]")
        val scc = new StreamingContext(conf, Seconds(5))
        val sc = scc.sparkContext
        // 创建一个可变队列
        val queue: mutable.Queue[RDD[Int]] = mutable.Queue[RDD[Int]]()
        val rddDS: InputDStream[Int] = scc.queueStream(queue, true)
        rddDS.reduce(_ + _).print
        scc.start
        // 循环的方式向队列中添加 RDD
        for (elem <- 1 to 5) {
            queue += sc.parallelize(1 to 100)
            Thread.sleep(2000)
        }
        scc.awaitTermination()
    }
}
```

6.3.2 自定义数据源

1. 使用及说明

其实就是自定义接收器，需要继承 Receiver，并实现 onStart、onStop 方法来自定义数据源采集。

2. 需求

自定义数据源，实现监控某个端口号，并获取该端口的内容。

3. 代码

首先自定义数据源，实现代码如下所示。

代码 6-3　MySource.scala

```
object MySource{
    def apply(host: String, port: Int): MySource = new MySource(host, port)
}
class MySource(host: String, port: Int) extends
Receiver[String](StorageLevel.MEMORY_ONLY){
    /*
    *接收器启动的时候调用该方法。
    *这个函数内部必须初始化一些读取数据必需的资源(threads、buffers 等)
    *该方法不能阻塞，所以读取数据要在一个新的线程中进行
    */
    override def onStart(): Unit = {
        // 启动一个新的线程来接收数据
        new Thread("Socket Receiver"){
            override def run(): Unit = {
                receive()
            }
        }.start()
    }
    // 此方法用来接收数据
```

```
    def receive()={
        val socket = new Socket(host, port)
        val reader = new BufferedReader(new
InputStreamReader(socket.getInputStream, StandardCharsets.UTF_8))
        var line: String = null
        // 当 receiver 没有关闭且 reader 读取到了数据时，循环发送给 Spark
        while (!isStopped && (line = reader.readLine()) != null ){
            // 发送给 Spark
            store(line)
        }
        // 循环结束，则关闭资源
        reader.close()
        socket.close()
        // 重启任务
        restart("Trying to connect again")
    }
    override def onStop(): Unit = {
    }
}
```

然后，使用自定义数据源：

代码 6-4 MySourceDemo.scala

```
object MySourceDemo {
    def main(args: Array[String]): Unit = {
        val conf = new
SparkConf().setAppName("StreamingWordCount").setMaster("local[*]")
        // 1. 创建 SparkStreaming 的入口对象——StreamingContext，第 2 个参数表示事件间隔
        val ssc = new StreamingContext(conf, Seconds(5))
        // 2. 创建一个 DStream
        val lines: ReceiverInputDStream[String] =
ssc.receiverStream[String](MySource("hadoop201", 9999))
        // 3. 一个一个的单词
        val words: DStream[String] = lines.flatMap(_.split("""\s+"""))
        // 4. 单词形成元组
        val wordAndOne: DStream[(String, Int)] = words.map((_, 1))
        // 5. 统计单词的个数
        val count: DStream[(String, Int)] = wordAndOne.reduceByKey(_ + _)
        //6. 显示
        count.print
        //7. 启动流式任务开始计算
        ssc.start()
        //8. 等待计算结束才退出主程序
        ssc.awaitTermination()
        ssc.stop(false)
    }
}
```

最后，开启端口：

```
nc -lk 10000
```

6.3.3 Kafka 数据源

1. 用法及说明

在工程中需要引入 Maven 依赖 spark-streaming-kafka_2.11 来使用它。包内提供的 KafkaUtils 对象可以在 StreamingContext 和 JavaStreamingContext 中用 Kafka 消息创建出 DStream。

两个核心类：KafkaUtils、KafkaCluster。

2. 导入依赖

```
<dependency>
    <groupId>org.apache.spark</groupId>
    <artifactId>spark-streaming-kafka-0-8_2.11</artifactId>
    <version>2.1.1</version>
</dependency>
```

3. 高级 API

代码 6-5　HighKafka .scala

```
import kafka.serializer.StringDecoder
import org.apache.kafka.clients.consumer.ConsumerConfig
import org.apache.spark.SparkConf
import org.apache.spark.streaming.kafka.KafkaUtils
import org.apache.spark.streaming.{Seconds, StreamingContext}
object HighKafka {
    def main(args: Array[String]): Unit = {
        val conf: SparkConf = new
SparkConf().setMaster("local[*]").setAppName("HighKafka")
        val ssc = new StreamingContext(conf, Seconds(3))
        // Kafka 参数
        //Kafka 参数声明，以下 brokers 参照 "5.4 Kafka 集群部置"
        val brokers = "hadoop201:9092,hadoop202:9092,hadoop203:9092"
        val topic = "first"
        val group = "bigdata"
        val deserialization =
"org.apache.kafka.common.serialization.StringDeserializer"
        val kafkaParams = Map(
            ConsumerConfig.GROUP_ID_CONFIG -> group,
            ConsumerConfig.BOOTSTRAP_SERVERS_CONFIG -> brokers,
        )
        val dStream = KafkaUtils.createDirectStream[String, String,
StringDecoder, StringDecoder](
            ssc, kafkaParams, Set(topic))
        dStream.print()
        ssc.start()
        ssc.awaitTermination()
    }
}
```

6.4　DStream 实战

本节进行 DStream 实战。主要分为四个部分，首先讲解 DStream 程序从端口读取数据到控制台，然后讲解 FileStream 从文件系统读取数据，然后讲解窗口函数计算连续的任意多个 DStream 中 RDD 的数据，最后讲解 updateStateByKey，用于计算从开始到目前接收到的所有数据的统计信息。

6.4.1　从端口读取数据

以下是一个 DStream 的示例程序，也是读取 9999 端口的数据并输出到控制台。

代码 6-6　DStreaming.scala

```scala
package org.hadoop.spark
import org.apache.spark.streaming.dstream.{DStream, ReceiverInputDStream}
import org.apache.spark.streaming.{Seconds, StreamingContext}
import org.apache.spark.{SparkConf, SparkContext}
object DStreaming {
  def main(args: Array[String]): Unit = {
    //配置 sparkConf 参数
    val sparkConf: SparkConf = new SparkConf()
      .setAppName("SparkStreamingTCP").setMaster("local[2]")
    //构建 sparkContext 对象
    val sc: SparkContext = new SparkContext(sparkConf)
    //设置日志输出级别
    sc.setLogLevel("WARN")
    //构建 StreamingContext 对象，每个批处理的时间间隔
    val scc: StreamingContext = new StreamingContext(sc, Seconds(5))
    //注册一个监听的 IP 地址和端口，用来收集数据
    val lines: ReceiverInputDStream[String] =
      scc.socketTextStream("192.168.56.201", 9999)
    //切分每一行记录
    val words: DStream[String] = lines.flatMap(_.split(" "))
    //每个单词记为 1
    val wordAndOne: DStream[(String, Int)] = words.map((_, 1))
    //分组聚合
    val result: DStream[(String, Int)] = wordAndOne.reduceByKey(_ + _)
    //打印数据
    result.print()
    scc.start()
    scc.awaitTermination()
  }
}
```

6.4.2　FileStream

FileStream 更准确地说应该叫作目录文件流，它的功能是用于监听一个目录下的文件增加，如果增加了一个文件，将会读取这个文件的内容并处理。注意，此类在 Windows 环境下无效，只能运

行在 Unix/Linux/Mac 环境下。

1. 监听 Linux 上的某个目录

代码 6-7　FileStreaming.scala

```scala
package org.hadoop.spark
import org.apache.spark.{SparkConf, SparkContext}
import org.apache.spark.streaming.dstream.DStream
import org.apache.spark.streaming.{Seconds, StreamingContext}
object FileStreaming {
  def main(args: Array[String]): Unit = {
    val conf = new SparkConf()
      .setMaster("local[2]") //至少两个线程
      .setAppName("FileStream")
    //声明 SparkStreaming 对象
    val ssc = new StreamingContext(conf, Seconds(2));
    val context = ssc.sparkContext
    context.setLogLevel("WARN");
    //监听这个目录下文件的增加，如果有文件增加，就会读取文件内容并处理
    val rdd: DStream[String] = ssc.textFileStream("file:///home/hadoop/1");
    rdd.flatMap(_.split("\\s+")).map((_, 1)).reduceByKey(_ + _).print();
    ssc.start();
    ssc.awaitTermination();
    Thread.sleep(1000*20);
    ssc.stop();
  }
}
```

2. 监听 HDFS 上某个目录下文件的增加

当监听 HDFS 上某个目录下文件的增加时，出现如图 6-11 所示的异常。

```
2019-01-20 15:32:20 ERROR JobScheduler:91 - Error generating jobs for time 1547969540000 ms
org.apache.hadoop.mapreduce.lib.input.InvalidInputException: Input path does not exist: hdfs://server21:8020/test/1.txt._COPYING_
    at org.apache.hadoop.mapreduce.lib.input.FileInputFormat.singleThreadedListStatus(FileInputFormat.java:325)
    at org.apache.hadoop.mapreduce.lib.input.FileInputFormat.listStatus(FileInputFormat.java:265)
```

图 6-11

原因是使用 HDFS 上传文件时，会先形成一个_COPYING_的文件，这个_COPYING_文件在文件上传完成以后会被删除，但它也会被 Spark Stream 监听，当处理时它已经被删除，所以出现上述错误。此时，就必须使用 fileStream，因为它拥有第二个参数 filter。

修改上述的代码，注意使用 fileStream 而不是 textFileStream 以避免图 6-11 所示的错误。

代码 6-8　FileStreaming2.scala

```scala
package org.hadoop.spark
import org.apache.hadoop.fs.Path
import org.apache.hadoop.io.{LongWritable, Text}
import org.apache.hadoop.mapreduce.lib.input.TextInputFormat
import org.apache.spark.SparkConf
import org.apache.spark.streaming.dstream.DStream
import org.apache.spark.streaming.{Seconds, StreamingContext}
```

```
object FIleStreaming2 {
  def main(args: Array[String]): Unit = {
    if (args.length != 1) {
      println("usage : <in>")
      return;
    }
    val conf = new SparkConf()
      .setAppName("FileStream")
    //声明 SparkStreaming 对象
    val ssc = new StreamingContext(conf, Seconds(2));
    val rdd: DStream[(LongWritable, Text)] =
      ssc.fileStream[LongWritable, Text, TextInputFormat](directory = args(0),
//目录
      filter = ((path: Path) => (!path.getName.contains("_COPYING_"))), //过
滤条件
      newFilesOnly = true);
    rdd.map(_._2.toString). //先获取 value 数据
      flatMap(_.split("\\s+")) //再对文本进行处理
      .map((_, 1)).reduceByKey(_ + _).print();
    ssc.start();
    ssc.awaitTermination();
    Thread.sleep(1000 * 20);
    ssc.stop();
  }
}
```

然后打包，再执行以下命令提交代码：

```
$spark-submit --master spark://server201:7077 \
--class org.hadoop.spark.FileStreaming2 \
chapter11-1.0.jar hdfs://server201:8020/test/
```

最后使用 HDFS 上传文件到 /test 目录下，查看 Steaming 的输出信息即可。

6.4.3 窗口函数

窗口函数 window 可以计算连续的任意多个 DStream 中 RDD 的数据，如图 6-12 所示。

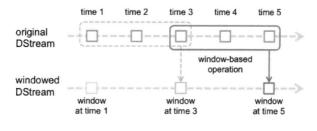

图 6-12

window 函数有两个主要的参数：

- 窗口长度 - 窗口的持续时间：窗口的个数。

- 滑动间隔 - 执行窗口操作的间隔：滑动的块数。

PairDStreamFunctions 拥有一个函数，用于处理窗口程序，这个函数的源码如下：

```
def reduceByKeyAndWindow(
    reduceFunc: (V, V) => V,
    windowDuration: Duration,
    slideDuration: Duration
  ): DStream[(K, V)] = ssc.withScope {
    reduceByKeyAndWindow(reduceFunc, windowDuration, slideDuration,
defaultPartitioner())
  }
```

现在我们开发一个 Java 客户端，每 2 秒向 9999 端口请求一次数据，设置窗口个数为 3，即 2×3=6，设置滑块个数为 2，即 2×2=4。现在我们在 Netcat 控制台输入数据并查看 Streaming 接收到的数据。

代码 6-9　WindowFun.scala

```scala
package org.hadoop.spark
import org.apache.spark.SparkConf
import org.apache.spark.storage.StorageLevel
import org.apache.spark.streaming.{Seconds, StreamingContext}
object WindowFun {
  def main(args: Array[String]): Unit = {
    val conf = new SparkConf()
      .setMaster("local[2]")
      .setAppName("WindowFun")
    //声明 SparkStreaming 对象
    val ssc = new StreamingContext(conf, Seconds(2));
    val context = ssc.sparkContext;
    context.setLogLevel("WARN");
    //必须要声明检查点
    ssc.checkpoint("file:///D:/a/log");
    val rdd = ssc.socketTextStream("server201", 9999,
StorageLevel.MEMORY_AND_DISK);
    val rdd2 =  rdd.flatMap(_.split("\\s+")).map((_,1))
      .reduceByKey(_+_).reduceByKeyAndWindow((a:Int,b:Int)=>a+b,Seconds(6),Seconds(4));
    rdd2.print();
    //开始
    ssc.start();
    ssc.awaitTermination();
  }
}
```

6.4.4　updateStateByKey

updateStateByKey 用于计算从开始到目前接收到的所有数据的统计信息。

为了可以均匀地向 Spark Streaming 传递 socket 数据，这里使用 Java 代码发送。代码如下：

代码 6-10　ForUpdateStateByKey.java

```java
package org.hadoop.spark;
import java.io.PrintStream;
import java.net.ServerSocket;
import java.net.Socket;
public class ForUpdateStateByKey {
    public static void main(String[] args) throws Exception {
        ServerSocket ss = new ServerSocket(9999);
        while (true) {
            System.out.println("等待连接...");
            Socket client = ss.accept();
            System.out.println("连接，启动线程..." + client);
            new MyThread(client).start();
        }
    }
    /**
     * 使用以下线程，向连接成功的客户端发送数据，第一秒发送一次相同的数据
     **/
    public static class MyThread extends Thread {
        private Socket client;
        public MyThread(Socket client) {
            this.client = client;
        }
        @Override
        public void run() {
            try {
                int times = 1;
                PrintStream ps = new PrintStream(client.getOutputStream(), true);
                while (true) {//每一秒向客户端发送一个信息
                    ps.println((times++) + " Jack Mary");
                    Thread.sleep(1000);
                }
            } catch (Exception e) {
                e.printStackTrace();
            }
        }
    }
}
```

Spark Streaming 客户端的代码如下：

代码 6-11　UpdateStateByKeyClient.scala

```scala
package org.hadoop.spark
import org.apache.spark.SparkConf
import org.apache.spark.streaming.{Seconds, StreamingContext}
object UpdateStateByKeyClient {
  def main(args: Array[String]): Unit = {
    val conf = new SparkConf()
      .setMaster("local[2]")
```

```
      .setAppName("NetworkWordCount")
    //声明 SparkStreaming 对象
    val ssc = new StreamingContext(conf, Seconds(1));
    //必须要设置检查点
    ssc.checkpoint("file:///D:/a/cp");
    //声明 updateStateByKey 函数=(参数){函数体}
    /** 其中第一个参数为前面累计的结果,第二个参数为新的结果 * */
    val updateStateByKey = (values: Seq[Int], state: Option[Int]) => {
      val newValue: Int = state.getOrElse(0);
      //如果没有则为 0
      //将前面的值进行累计求和
      val sum: Int = values.sum;
      Some(sum + newValue);  //这是返回值
    };
    val rdd = ssc.socketTextStream("server201", 9999);
    //转换完成以后再执行 update 函数
    val rdd2 = rdd.flatMap(_.split("\\s+")).map((_, 1))
      .updateStateByKey[Int](updateStateByKey);
    rdd2.print();
    ssc.start();
    ssc.awaitTermination();
    Thread.sleep(1000 * 30);
    ssc.stop();
  }
}
```

先启动 Java 代码,再启动 Scala 代码,查看 Scala 客户端 Spark Streaming 接收到的数据:

```
-------------------------------------------
Time: 1547973415000 ms   第一次连接,显示 Mary 和 Jack 都是 1 个
-------------------------------------------
(Mary,1)
(1,1)
(Jack,1)

-------------------------------------------
Time: 1547973416000 ms 第二个获取信息,显示 Mary 和 Jack 都是 2 个,后面以此类推
-------------------------------------------
(2,1)
(Mary,2)
(1,1)
(Jack,2)
-------------------------------------------
Time: 1547973417000 ms
-------------------------------------------
(2,1)
(3,1)
(Mary,3)
(1,1)
(Jack,3)
```

6.5 Structured Streaming

本节主要介绍 Structured Streaming 的相关内容，包括 Structured Streaming 概述及其快速示例。

6.5.1 概述

Structured Streaming 是一个可扩展、容错的流处理引擎，建立在 Spark SQL 引擎之上。开发者可以用与离线批处理数据相同的方式来表示流计算的逻辑，并且保持其逻辑的一致性（流批一体）。Spark SQL 引擎会处理好增量以连续运行，并随着流式数据的接收持续更新最终结果。开发者可以使用 Dataset/DataFrame API 通过 Scala、Java、Python 或者 R 语言编程来表达 streaming 聚合、事件时间窗口、流批 Join 等。计算逻辑在 Spark SQL 引擎上执行，充分利用了 Spark SQL 引擎的优势。最后，系统通过 Checkpoint 及 Write-Ahead Log 保证端到端的 exactly-once 容错机制。简单来说，Structured Streaming 提供了高性能、可扩展、容错、端到端 exactly-once 的流处理。下一小节是一个 DStream 的快速示例程序，同样也是读取 9999 端口的数据并输出到控制台。

6.5.2 快速示例

假设我们需要维护一个持续从 TCP socket 接收文本数据并进行单词计数（word count）的程序，那么应该如何使用 Structured Streaming 来表达呢？下面是 Scala 的例子，让我们通过示例一步一步地理解其工作原理。如果搭建好 Spark 环境，还可以直接运行该示例。

首先，我们需要导入必需的类并创建一个本地模式的 SparkSession，这是实现所有与 Spark 相关功能的起点。

```
import org.apache.spark.sql.functions._
import org.apache.spark.sql.SparkSession
val spark = SparkSession
  .builder
  .appName("StructuredNetworkWordCount")
  .getOrCreate()
import spark.implicits._
```

接下来，创建一个 Streaming DataFrame，它表示从本地主机 9999 端口接收的文本数据，然后转换 DataFrame 来计算 WordCount。

```
// 创建 DataFrame，代表连接到 localhost:9999 的输入流
val lines = spark.readStream
  .format("socket")
  .option("host", "localhost")
  .option("port", 9999)
  .load()
// 将语句分割为单词
val words = lines.as[String].flatMap(_.split(" "))
// 生成运行 WordCount 的 DataFrame
val wordCounts = words.groupBy("value").count()
```

lines DataFrame 表示包含流式文本数据的无界输入表，此表包含一列名为"value"的字符串，

流式文本数据中的每一行都成为表中的一行。注意，由于我们只是在设置转换，还没有开始转换，所以目前还没有接收到任何数据。接下来，我们将 DataFrame 转换为字符串数据集作为[String]，这样就可以应用 flatMap 操作将每一行拆分为多个单词。结果单词数据集包含所有单词。最后，我们通过按数据集中的唯一值分组并计数来定义 wordCounts DataFrame。注意，这是一个流 DataFrame，用于表示流的 WordCount。

我们现在已经设置了对流数据的查询（query），剩下的就是接收数据并统计计数。为了做到这一点，我们设置为在每次更新时将完整的计数集（由 outputMode("complete")指定）打印到控制台，然后使用 start()启动流计算。

```
// 开始运行查询，将运行中的计数打印到控制台
val query = wordCounts.writeStream
  .outputMode("complete")
  .format("console")
  .start()
query.awaitTermination()
```

执行此代码后，流计算将在后台启动。query 对象是活动流式查询的句柄，我们使用 waitTermination()等待查询的终止，以防止进程在查询活动时退出。

要实际执行此示例代码，我们可以在自己的 Spark 应用程序中编译代码，或者在下载 Spark 后运行示例。下面展示如何在 Spark 中运行示例代码。首先，我们需要启动 9999 端口：

```
$ nc -lk 9999
```

然后，在另一个终端中观察结果，使用以下代码，在运行 Netcat 服务器的终端中输入的任何行，都将每秒计数一次并打印在屏幕上。

```
$ ./bin/run-example org.apache.spark.examples.sql.streaming.StructuredNetworkWordCount localhost 9999
```

结果如下：

```
# TERMINAL 1:
# Running Netcat
$ nc -lk 9999
apache spark
apache hadoop
# TERMINAL 2: RUNNING StructuredNetworkWordCount
$ ./bin/run-example
org.apache.spark.examples.sql.streaming.StructuredNetworkWordCount localhost 9999
-------------------------------------------
Batch: 0
-------------------------------------------
+------+-----+
| value|count|
+------+-----+
|apache|    1|
| spark|    1|
+------+-----+
```

Structured Streaming 的关键思想是将实时数据流视为一个不断追加的表。这种思想创造了一种新的流处理模型,它与批处理模型非常相似。我们把流计算表示为静态表上的标准批处理查询,Spark 将它作为无界输入表上的增量查询来运行。下面详细介绍 Structured Streaming。

将输入数据流视为"输入表",流中的每个数据项都作为一个新行被追加到输入表中,如图 6-13 所示。

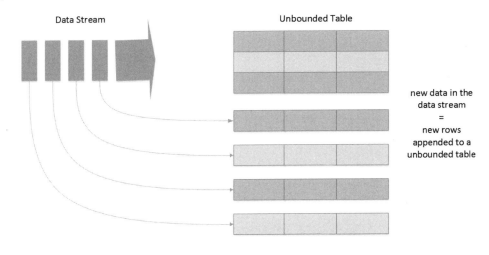

图 6-13

对输入的查询将生成"结果表",每个触发间隔(比如,每 1 秒)都会向输入表追加新行,最终更新结果表。无论何时更新结果表,我们都希望将更改后的结果行写入外部接收器。

"Output"定义为写入外部存储器的内容,可以在不同的模式下定义输出:

- Complete Mode:整个更新的结果表将写入外部存储器,由存储连接器决定如何处理整个表的写入。
- Append Mode:只有自上次触发以来追加到结果表中的新行才会写入外部存储器。这仅适用于结果表中的现有行预计不会更改的查询。
- Update Mode:只有自上次触发后在结果表中更新的行才会写入外部存储器(自 Spark 2.1.1 起可用)。注意,这与 Complete Mode 的不同之处在于此模式仅输出自上次触发以来已更改的行。如果查询不包含聚合,则相当于追加模式。

注意，每种模式都适用于某些类型的查询。

为了说明该模型的使用，让我们结合上面创建 Streaming DataFrame 的快速示例来理解该模型。示例中第一行 DataFrame 是输入表，最后一行 wordCounts DataFrame 是结果表。注意，从 Streaming DataFrame 的 lines 到 wordCounts 的 query，它们与 Static DataFrame 完全相同。但是，当该查询启动时，Spark 将持续检查来自 socket 连接的新数据。如果有新数据，那么 Spark 将运行"增量"查询，将以前运行的计数与新数据结合起来计算更新的计数。

总之，Structured Streaming 可以理解为结构化流，相当于把 spark sql 实时化了，底层基于 spark sql。对无边界，无序的数据源，允许按数据本身的特征进行窗口计算，得到基于事件发生时间的有序结果，并能在准确性、延迟程度和处理成本之间调整。Structured Streaming 统一了流式计算和批处理的 API，都是使用 Data frame/ Data Set，特别适用于结构化的数据。

第 7 章

Spark ML 机器学习

人工智能时代的到来，主流科技企业都在积极地将自己的发展方向重定位于围绕人工智能（AI）和机器学习（ML）方向。机器学习是一门多领域交叉学科，涉及概率论、统计学、逼近论、凸分析、算法复杂度理论等多门学科，专门研究计算机怎样模拟或实现人类的学习行为，以获取新的知识或技能，重新组织已有的知识结构来不断改善自身的性能。大部分 AI 应用背后的真正驱动力是传统的机器学习模型，而不是深度神经网络。工程师们使用传统的软件和工具来实现机器学习工程，然后发现这些软件和工具并不是很有效：将数据传输到模型并最终输出结果的整个流水线，最终只能产生一些零散的、互不兼容的东西。Spark ML 是 Spark 提供的处理机器学习方面的功能库，该库包含了许多机器学习算法，开发者不需要深入了解机器学习算法就能开发出相关程序。本章将介绍 Spark ML 库的基本知识及其使用方法。

本章主要知识点：

- 机器学习概述
- Spark ML 介绍
- 典型的机器学习流程
- 典型的算法模型实战

7.1 机 器 学 习

机器学习是一类算法的总称，这些算法企图从大量历史数据中挖掘出其中隐含的规律，并用于预测或者分类。更具体地说，机器学习可以看作寻找一个函数，输入是样本数据，输出是期望的结果，只是这个函数过于复杂，以至于不太方便形式化表达。需要注意的是，机器学习的目标是使学到的函数很好地适用于"新样本"，而不仅仅是在训练样本上表现良好。学到的函数适用于新样本的能力，称为泛化（Generalization）能力。

机器学习的一般步骤如下：

步骤01 选择一个合适的模型。这通常需要依据实际问题而定,针对不同的问题和任务需要选取恰当的模型,模型就是一组函数的集合。

步骤02 判断一个函数的好坏。这需要确定一个衡量标准,也就是我们通常说的损失函数(Loss Function),损失函数的确定也需要依据具体问题而定,比如,回归问题一般采用欧式距离,分类问题一般采用交叉熵代价函数。

步骤03 找出"最好"的函数。如何从众多函数中最快地找出"最好"的那一个,这是最大的难点,做到又快又准往往不是一件容易的事情。常用的方法有梯度下降算法、最小二乘法等,以及其他一些技巧(tricks)。

学习得到"最好"的函数后,需要在新样本上进行测试,只有在新样本上也表现良好,才算是一个"好"的函数。

基于它与经验、环境或者任何我们称之为输入数据的相互作用,一个算法可以用不同的方式对一个问题建模。流行的做法是首先考虑一个算法的学习方式。算法的主要学习方式和学习模型只有几个,下面我们将逐一介绍它们,并且给出几个算法和它们适合解决的问题类型来作为例子。

- 监督学习:输入数据被称为训练数据,它们有已知的标签或者结果,比如垃圾邮件/非垃圾邮件或者某段时间的股票价格。模型的参数确定需要通过一个训练的过程,在这个过程中模型将会做出预测,当预测不符时,则需要做出修改。
- 无监督学习:输入数据不带标签或者没有一个已知的结果。通过推测输入数据中存在的结构来建立模型。这类问题的例子有关联规则学习和聚类。算法的例子有 Apriori 算法和 K-means 算法。
- 半监督学习:输入数据由带标记的和不带标记的数据组成。合适的预测模型虽然已经存在,但是模型在预测的同时还必须通过发现潜在的结构来组织数据。这类问题包括分类和回归。典型算法包括对一些其他灵活模型的推广,这些模型都对如何给未标记数据建模做出了一些假设。
- 强化学习:输入数据作为来自环境的激励提供给模型,且模型必须做出反应。反馈并不像监督学习那样来自训练的过程,而是作为环境的惩罚或者奖赏。典型问题有系统和机器人控制。算法的例子包括 Q-学习和时序差分学习(Temporal Difference Learning)。当我们处理大量数据来对商业决策建模时,通常会使用监督学习和无监督学习。

目前的一个热门话题是半监督学习,比如在图像分类中,涉及的数据集很大但是只包含极少数标记的数据。通常我们会把算法按照功能和形式的相似性来进行区分,比如树形结构和神经网络的方法。这是一种有用的分类方法,但也不是完美的,仍然有些算法很容易就可以被归入好几个类别,比如学习矢量量化,它既是受启发于神经网络的方法,又是基于实例的方法;也有一些算法的名字既描述了它处理的问题,也是某一类算法的名称,比如回归和聚类。正因如此,我们会从不同的来源看到对算法进行的不同归类。就像机器学习算法自身一样,没有完美的模型,只有足够好的模型。

随着大数据的发展,人们对大数据的处理要求也越来越高,原有的批处理框架 MapReduce 适合离线计算(在 MapReduce 中,由于其分布式特性,所有数据需要读写磁盘,启动工作耗时较大,难以满足时效性要求),却无法满足实时性要求较高的业务,如实时推荐、用户行为分析等。

Spark 是一个类似于 MapReduce 的分布式计算框架，其核心是弹性分布式数据集，提供了比 MapReduce 更丰富的模型，可以快速在内存中对数据集进行多次迭代，以支持复杂的数据挖掘算法和图形计算算法。

7.2 Spark ML

对于机器学习，Spark 3.0 之后推荐使用 ML 库，而 MLlib 进入维护状态，不建议使用。ML 主要操作的是 DataFrame，而 MLlib 操作的是 RDD，相比于 MLlib 在 RDD 上提供的基础操作，ML 在 DataFrame 上的抽象级别更高，数据和操作耦合度更低。

Spark ML 是 Spark 3.0 版本的机器学习库，提供了基于 DataFrame 高层次的 API，可以用来构建机器学习管道。之所以使用 Spark ML，是因为基于 DataFrames 的 API 更加通用且灵活。

常见的机器学习问题有 4 种分类、回归、聚类和协同过滤。常见的算法类和工具类如图 7-1 所示。

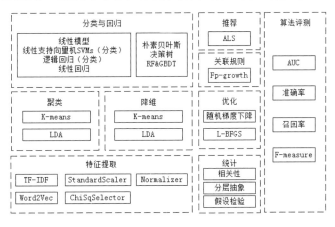

图 7-1

7.3 典型机器学习流程介绍

典型机器学习流程是"提出问题→假设函数→确定损失函数→训练模型确定参数"。

7.3.1 提出问题

图 7-2（左）所示是北京市海淀区的房价及面积关系数据，由这些数据估计出海淀区某个房子的售价大概在多少万元。图 7-2（右）是房价及面积关系图，在图中用一些离散的点把这些数据表示出来，横坐标是面积，纵坐标是售价。从图中可以看到数据存在一定的规律，比如 200 平方米的房屋售价上千万了。

机器学习就是用来做这样的事情的，也就是从已知的数据中找出规律，方便做出预测或者估计。那如何去找出规律呢？用直线去拟合离散点，如图 7-3 所示。

售价(万元)	建筑面积	室	厅
825.0	135.00	3	2
997.5	133.00	3	2
1005.0	134.00	3	2
384.00	64.00	3	2
270.00	45.00	2	1
459.6	76.00	1	1
388.8	64.8	1	1
713.4	118.9	1	1
218.4	39.0	1	1
1145.5	145.0	3	2
1864.4	236.00	4	2
539.0	77.00	2	2
679.0	97.00	3	2
756.00	108.00	3	2
784.00	112.00	3	2
487.5	75.0	1	2
780.0	120.00	2	2
780.0	120.00	2	2
780.0	120.00	3	2
995.5	147.0	3	2
1072.5	165.00	3	2
1072.5	165.00	3	2

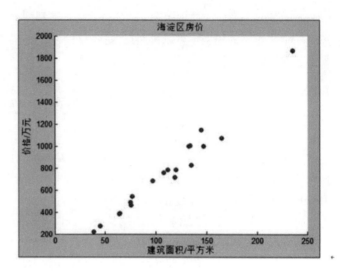

图 7-2

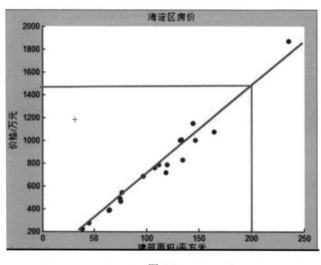

图 7-3

从图 7-3 中可以看到，图中的点几乎都围绕在直线附近，从而可以看出 200 平方米的房子的售价基本上在 1400 万到 1600 万之间。机器学习就是要寻找这样的直线，从而找到房价的规律。

7.3.2 假设函数

通过 7.3.1 节我们初步判定样本点分布在一条直线的周边，可以考虑采用线性模型进行求解，首先就是要设定一个假设函数。平面几何里的这条直线用函数表示为：$h(x) = \theta_0 x_0 + \theta_1 x_1$。这种函数我们把它叫作线性函数。图 7-3 中的这条直线用公式表示为：房价=0.0+θ_1×面积。我们尝试绘制三条直线，其中 $\theta_0=0$，θ_1 可以取值 6、7、8，不同参数对应的直线如图 7-4 所示。

θ_0、θ_1 可以有更多的取值，而当它们取不同值时，它们与真实数据的吻合程度也各不相同。比如最下面那条直线（$\theta_1=6$），可以看到位置和样本点之间还是有所偏差的，而 $\theta_1=7$ 时大部分房价是吻合的，那我们就可以以 $\theta_0=0$ 和 $\theta_1=7$ 来预测房价了。

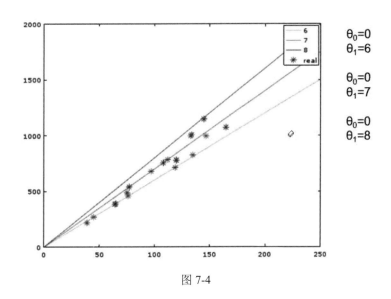

图 7-4

θ_0、θ_1 可以有更多的取值,取不同值时,它们与真实数据的吻合程度不同。比如图 7-4 中最下面那条直线($\theta_1=6$),可以看到位置和样本点之间还是有所偏差的,而 $\theta_1=7$ 时大部分房价是吻合的,那我们就可以以 $\theta_0=0$ 和 $\theta_1=7$ 来预测房价了。我们通过观察曲线就能知道 $\theta_1=7$ 最好,但计算机怎么知道呢?答案就是通过计算损失函数获得。

7.3.3 损失函数

所谓损失价函数,也叫代价函数,用来判断预测函数 $h(x)$ 和真实数据之间的误差程度。不同直线的误差程度如图 7-5 所示。

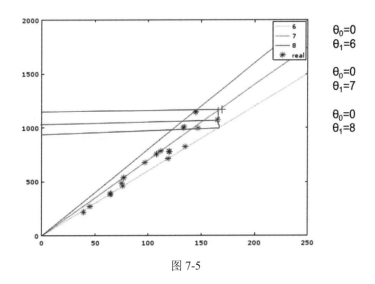

图 7-5

将样本点的横坐标代入三个不同的直线函数 $h(x)$,计算出房价,然后和真实房价做比较,可以发现最下面那条直线的值更加接近真实房价。

通过上面的分析,问题变化为寻找一个最接近真实规律的 θ_0 和 θ_1,最终我们就得到了下面的损失函数。损失函数有很多种表现方式，这个是一个比较好理解的方式。

$$J(\theta_0, \theta_1) = \frac{1}{2m} \sum (h(x_{(m,2)}) - y_{(m,1)})^2$$

参数说明如下：

- h 是预测函数。
- x 是面积特征，包括 θ_0 和 θ_1 两个特征。
- y 是房价标签，是真实的数据。
- （预测值-真实值）^{2}，因为误差比较小，所以放大这个误差，取平方值。
- Σ 是求和函数，把所有差值的平方求和然后求平均。
- m 就是真实样本数据的数量，如果我们有 100 条数据，那么 m=100。

这个损失函数用来衡量预测值 $h(x_i)$ 与真实样本数据 y_i 之间的差距。

假设我们给定了 θ_0 和 θ_1，那么这个结果值是大了好，还是小了好呢？当然是越小越好，越小说明预测值和真实值越接近。

7.3.4 训练模型确定参数

上一小节中给出了损失函数，接下来就是计算 θ_0 和 θ_1 了。有效的办法就是不断尝试改变 θ_0 和 θ_1。先尝试计算出 θ_0 和 θ_1 都是 0 时的结果，然后 θ_0 改成 1，与刚才的结果比较看哪个比较小，如果 θ_0=1 的结果较小那当然好，如果 θ_0=1 的结果较大，就把这组结果排除，再试下一组。总之就是不断地尝试 θ_0 和 θ_1，让计算的结果最小。

上面这套规则以及样本数据就被称为模型。不断尝试改变 θ_0 和 θ_1 的值来求最合适的 θ 的过程被称为训练模型。

7.4 经典算法模型实战

本章涉及的 Spark 机器学习的代码均在 Spark 本地模式上验证，这种模式方便读者进行代码的快速开发和调试，详细源码（可在本书的配套资源中获取），采用的是 Spark 3.0 版本的机器学习库 spark.ml 包。

本地运行模式也被称为 Local[N]模式，是用单机的多个线程来模拟 Spark 分布式计算，直接运行在本地电脑，便于调试，通常用来验证开发出来的应用程序在逻辑上有没有问题。其中 N 代表可以使用 N 个线程，每个线程拥有一个 core。如果不指定 N，则默认是 1 个线程（该线程有 1 个 core）。

7.4.1 聚类算法实战

1. 算法知识点

聚类（Cluster analysis）有时也被翻译为簇类，其核心任务是将一组目标 object 划分为若干个簇，每个簇之间的 object 尽可能相似，簇与簇之间的 object 尽可能相异。聚类算法是机器学习（或者说是数据挖掘更合适）中重要的一部分，除了最为简单的 K-Means 聚类算法外，比较常见的还有层次法（CURE、CHAMELEON 等）、网格算法（STING、WaveCluster 等），等等。

这里给出一个较权威的聚类问题定义：所谓聚类问题，就是给定一个元素集合 D，其中每个元素具有 n 个可观察属性，使用某种算法将 D 划分成 k 个子集，要求每个子集内部的元素之间相异度尽可能低，而不同子集的元素相异度尽可能高。其中每个子集叫作一个簇。

聚类与分类不同：分类是示例式学习，要求分类前明确各个类别，并将每个元素映射到一个类别；而聚类是观察式学习，在聚类前可以不知道类别甚至不给定类别数量，是无监督学习的一种。目前聚类广泛应用于统计学、生物学、数据库技术和市场营销等领域，相应的算法也非常多。

K-means 聚类属于无监督学习，以往的回归、朴素贝叶斯、SVM 等都是有类别标签 y 的，也就是说已经给出了样例的分类，而聚类的样本中却没有给定 y，只有特征 x。聚类的目的是找到每个样本 x 潜在的类别 y，并将同类别 y 的样本 x 放在一起。比如，假设宇宙中的星星可以表示成三维空间中的点集(x,y,z)，聚类后的结果是一个一个的星团，星团里面的星星距离比较近，星团间的星星距离就比较远了。

K-means 算法的工作过程说明如下：首先从 n 个数据对象中任意选择 k 个对象作为初始聚类中心；然后对剩下的对象，根据它们与这些聚类中心的相似度（距离）分配给与它们最相似的（聚类中心所代表的）聚类；再计算每个所获新聚类的聚类中心（该聚类中所有对象的均值）；不断重复这一过程直到标准测度函数开始收敛为止。一般都采用均方差作为标准测度函数。

K-means 聚类具有以下特点：各聚类本身尽可能紧凑，而各聚类之间尽可能分开。

衡量样本点到聚类中心的相似度一般是基于距离方式进行的。欧几里得相似度计算是一种基于样本点之间的直线距离的计算方式。在相似度计算中，可以将不同的样本点定义为不同的坐标点，而特定目标定位坐标原点。使用欧几里得距离计算两个点之间的绝对距离，公式如下：

$$d = \sqrt{(x_1 - x_2)^2 + (y_1 - y_2)^2}$$

ML 中 K-means 在进行工作时设定了最大的迭代次数，因此在运行时达到设定的最大迭代次数就停止迭代。

K-means 由于算法设计的一些基本理念，在对数据进行处理时效率不高。ML 充分利用了 Spark 框架的分布式计算的便捷性，设计了一个包含 K-means++方法的并行化变体，称为 K-means||，从而提高了运算效率。

K-Means 算法的结果好坏依赖于对初始聚类中心的选择，容易陷入局部最优解，对 K 值的选择没有准则可依循，对异常数据较为敏感，只能处理数值属性的数据，聚类结构可能不平衡。

2. 案例说明

在本案例中将介绍 K-means 算法，K-means 属于基于平方误差的迭代重分配聚类算法，其核心思想十分简单，基本步骤如下：

步骤01 随机选择 K 个中心点。
步骤02 计算所有点到这 K 个中心点的距离，选择距离最近的中心点为其所在的簇。
步骤03 简单地采用算术平均数（mean）来重新计算 K 个簇的中心。
步骤04 重复 **步骤02** 和 **步骤03**，直至簇类不再发生变化或者达到最大迭代值。

步骤 05 输出结果。

3. 数据处理及算法应用

本案例使用的数据为 sample_kmeans_data.txt，可以在配套资源下载包中本项目根目录的 data/ 目录下找到。在该文件中提供了 6 个点的空间位置坐标，使用 K-means 算法对这些点进行分类。

本项目使用的 sample_kmeans_data.txt 文件的数据如下：

```
0.0 0.0 0.0
0.1 0.1 0.1
0.2 0.2 0.2
9.0 9.0 9.0
9.1 9.1 9.1
9.2 9.2 9.2
```

其中每一行都是一个坐标点的坐标值。

fit 方法是 ML 中 K-means 模型的训练方法，其内容如下：

```
Class KMeans extends Estimator[KMeansModel] with KMeansParams with DefaultParamsWritable
//KMeans 类
def fit(dataset: Dataset[_]): KMeansModel
//训练的方法
```

若干个参数可由一系列 setter 函数来设置，参数说明如下：

- data: Dataset[_]：输入的数据集。
- setK(value: Int)：聚类分成的数据集数。
- setMaxIter(value: Int)：最大迭代次数。

聚类算法的应用代码如代码 7-1 所示。

代码 7-1　KMeansExample.scala

```scala
import org.apache.spark.ml.clustering.KMeans
import org.apache.spark.ml.evaluation.ClusteringEvaluator
import org.apache.spark.sql.SparkSession
object KMeansExample {
  def main(args: Array[String]): Unit = {
    val spark = SparkSession
      .builder                          //创建 Spark 会话
      .master("local")                  //设置本地模式
      .appName("K-means")               //设置名称
      .getOrCreate()                    //创建会话变量
    //读取数据
    val dataset = spark.read.format("libsvm").load("data/sample_kmeans_data.txt")
    //训练模型，设置参数，载入训练集数据正式训练模型
    val kmeans = new KMeans().setK(3).setSeed(1L)
    val model = kmeans.fit(dataset)
    //使用测试集做预测
```

```
    val predictions = model.transform(dataset)
    //使用轮廓分数评估模型
    val evaluator = new ClusteringEvaluator()
    val silhouette = evaluator.evaluate(predictions)
    println(s"Silhouette with squared euclidean distance = $silhouette")
    //展示结果
    println("Cluster Centers: ")
    model.clusterCenters.foreach(println)
    spark.stop()
  }
}
```

其中项目中需要引入 Spark 核心包、机器学习包等依赖,本章项目对应的依赖的具体代码如下:

代码 7-2   pom.xml

```xml
<dependencies>
    <dependency>
        <groupId>org.scala-lang</groupId>
        <artifactId>scala-library</artifactId>
        <version>2.12.7</version>
    </dependency>
    <dependency>
        <groupId>org.apache.spark</groupId>
        <artifactId>spark-core_2.12</artifactId>
        <version>3.3.1</version>
    </dependency>
    <!--引入 sparkStreaming 依赖-->
    <dependency>
        <groupId>org.apache.spark</groupId>
        <artifactId>spark-streaming_2.12</artifactId>
        <version>3.3.1</version>
    </dependency>
    <!--引入 sparkstreaming 整合 Kafka 的依赖-->
    <dependency>
        <groupId>org.apache.spark</groupId>
        <artifactId>spark-streaming-kafka-0-8_2.11</artifactId>
        <version>2.0.2</version>
    </dependency>
    <dependency>
        <groupId>org.apache.spark</groupId>
        <artifactId>spark-mllib_2.12</artifactId>
        <version>3.3.1</version>
    </dependency>
</dependencies>
```

其中轮廓分数使用 ClusteringEvaluator,它测量一个簇中的每个点与相邻簇中的点的接近程度,从而帮助判断簇是否紧凑且间隔良好。时间复杂度为 $O(tknm)$,其中 t 为迭代次数、k 为簇的数目、n 为样本点数、m 为样本点维度。空间复杂度为 $O(m(n+k))$,其中 k 为簇的数目、m 为样本点维度、n 为样本点数。K-means 是对三维数据进行聚类处理,如果是更高维的数据,读者可自行修改数据集进行计算和验证,并自行打印验证程序的运行结果。

7.4.2 回归算法实战

1. 算法知识点

线性回归(Linear Regression)是利用被称为线性回归方程的函数,对一个或多个自变量和因变量之间的关系进行建模的一种回归分析方法。只有一个自变量的情况称为简单回归,大于一个自变量的情况叫作多元回归,在实际情况中大多数都是多元回归。

线性回归问题属于监督学习范畴,又被称为分类或归纳学习(Inductive Learning)。这类分析中训练数据集中给出的数据类型是确定的。机器学习的目标是对于给定的一个训练数据集,通过不断地分析和学习产生一个联系属性集合和类标集合的分类函数(Classification Function)或预测函数(Prediction Function),这个函数称为分类模型(Classification Model)或预测模型(Prediction Model)。通过学习得到的模型可以是一个决策树、规格集、贝叶斯模型或一个超平面。通过这个模型可以对输入对象的特征向量进行预测或对对象的类标进行分类。

回归问题中通常使用最小二乘法(Least Squares)来迭代最优的特征中每个属性的比重,通过损失函数或错误函数(Error Function)来设置收敛状态,即作为梯度下降算法的逼近参数因子。

2. 案例说明

线性回归分析的整个过程可以简单描述为如下 3 个步骤:

步骤01 寻找合适的预测函数,即 7.4.1 节中的 $h(x)$,用来预测输入数据的判断结果。这个过程非常关键,需要对数据有一定的了解或分析,知道或者猜测预测函数的大概形式,比如是线性函数还是非线性函数,若是非线性的,则无法用线性回归来得出高质量的结果。

步骤02 构造一个损失函数,该函数表示预测的输出(h)与训练数据标签(y)之间的偏差,可以是二者之间的差(h-y)或者是其他形式(如平方差、开方等)。综合考虑所有训练数据的"损失",将损失求和或者求平均,记为 $J(\theta)$ 函数,表示所有训练数据预测值与实际类别的偏差。

步骤03 显然,$J(\theta)$ 函数的值越小表示预测函数越准确(即 h 函数越准确),所以这一步需要做的是找到 $J(\theta)$ 函数的最小值。找函数的最小值有不同的方法,Spark 中采用的是梯度下降法。

3. 数据处理及算法应用

首先需要完成线性回归的数据准备工作。在 ML 中,线性回归的示例用来演示训练弹性网络(ElasticNet)正则化线性回归模型、提取模型汇总统计信息,以及使用 ElasticNet 回归综合。学习目标是最小化指定的损失函数,并进行正则化。

线性回归是进行连续值预测,最终获得一个基于特征变量的连续函数作为预测模型,衡量模型的准确度一般采用均方根误差(RMSE)。预测值和真实值的均方差越小,说明模型预测效果越好。具体代码如下:

代码 7-3 LinearRegression.scala

```
import org.apache.spark.ml.regression.LinearRegression
import org.apache.spark.sql.SparkSession
object LinearRegressionWithElasticNetExample {
```

```scala
    def main(args: Array[String]): Unit = {
      val spark = SparkSession
        .builder                                                    //创建 Spark 会话
        .master("local")                                            //设置本地模式
        .appName("LinearRegressionWithElasticNetExample")           //设置名称
        .getOrCreate()                                              //创建会话变量
      //$example on$
      //读取数据
      val training = spark.read.format("libsvm")
        .load("data/sample_linear_regression_data.txt")
      //建立一个 Estimator,并设置参数
      val lr = new LinearRegression()
        .setMaxIter(10)
        .setRegParam(0.3)  //正则化参数
        .setElasticNetParam(0.8)  //使用 ElasticNet 回归
      //训练模型
      val lrModel = lr.fit(training)
      //打印一些系数(回归系数表)和截距
      println(s"Coefficients: ${lrModel.coefficients} Intercept: ${lrModel.intercept}")
      //汇总一些指标并打印结果和一些监控信息
      val trainingSummary = lrModel.summary
      println(s"numIterations: ${trainingSummary.totalIterations}")
      println(s"objectiveHistory: [${trainingSummary.objectiveHistory.mkString(",")}]")
      trainingSummary.residuals.show()
      println(s"RMSE: ${trainingSummary.rootMeanSquaredError}")
      println(s"r2: ${trainingSummary.r2}")
      spark.stop()
    }
  }
```

其中,setElasticNetParam 设置的是 elasticNetParam,范围是 0~1,包括 0 和 1。如果设置的是 0,那么惩罚项是 L2 的惩罚项,训练的模型简化为 Ridge 回归模型;如果设置的是 1,那么惩罚项就是 L1 的惩罚项,等价于 Lasso 模型。

回归结果如下:

```
Coefficients:
[0.0,0.32292516677405936,-0.3438548034562218,1.9156017023458414,0.05288058680386263,0.765962720459771,0.0,-0.15105392669186682,-0.21587930360904642,0.22025369188813426] Intercept: 0.1598936844239736
  numIterations: 7
  objectiveHistory:
[0.4999999999999994,0.4967620357443381,0.4936361664340463,0.4936351537897608,0.4936351214177871,0.49363512062528014,0.4936351206216114]
+--------------------+
|           residuals|  (残差)
+--------------------+
|  -9.889232683103197|
|  0.5533794340053554|
```

```
|  -5.204019455758823|
| -20.566686715507508|
|   -9.4497405180564|
|  -6.909112502719486|
| -10.00431602969873|
|   2.062397807050484|
|   3.1117508432954772|
| -15.893608229419382|
|  -5.036284254673026|
|   6.483215876994333|
|  12.429497299109002|
| -20.32003219007654|
|  -2.0049838218725005|
| -17.867901734183793|
|   7.646455887420495|
|  -2.2653482182417406|
|  -0.10308920436195645|
|  -1.380034070385301|
+--------------------+
only showing top 20 rows

RMSE: 10.189077167598475
r2: 0.0228614669139581844
```

结果中，r2 表示的是判定系数，也称为拟合优度，其值越接近 1 越好。

7.4.3 协同过滤算法实战

1. 算法说明

协同过滤（Collaborative Filtering，CF）在 WIKI（维基百科）上的定义：简单来说是利用某个兴趣相投、拥有共同经验之群体的喜好来推荐感兴趣的资讯给使用者，个人通过合作的机制给予资讯相当程度的回应（如评分）并记录下来以达到过滤的目的，进而帮助别人筛选资讯，其中回应不一定局限于特别感兴趣的，特别不感兴趣资讯的记录也相当重要。

协同过滤常被应用于推荐系统。这些技术旨在补充用户—商品关联矩阵中所缺失的部分。

基于模型的协同过滤，用户和商品通过一小组隐性因子进行表达，并且这些因子也用于预测缺失的元素。ML 使用交替最小二乘法（ALS）来学习这些隐性因子。

这里对最小二乘法 ALS 做一下说明。ALS 是 Alternating Least Squares 的缩写，意为交替最小二乘法；而 ALS-WR 是 Alternating-Least-Squares with Weighted-λ-Regularization 的缩写，意为加权正则化交替最小二乘法，该方法常用于基于矩阵分解的推荐系统中。比如用户对商品的评分矩阵可以分解为一个用户对隐含特征偏好的矩阵和一个商品所包含的隐含特征的矩阵，即对于 $R(m \times n)$ 的矩阵，ALS 旨在找到两个低维矩阵 $X(m \times k)$ 和矩阵 $Y(n \times k)$，来近似逼近 $R(m \times n)$，在这过程中把用户评分缺失项填上，并根据这个分数对用户进行推荐。

根据应用本身的不同，用户对物品或者信息的偏好可能包括用户对物品的评分、用户查看物品的记录、用户的购买记录等。其实这些用户的偏好信息可以分为两类：

- 显式的用户反馈：这类是用户在网站上自然浏览或者使用网站以外的方式显式地提供的反馈信息，例如用户对物品的评分或者对物品的评论。
- 隐式的用户反馈：这类是用户在使用网站时产生的数据，隐式地反映了用户对物品的喜好，例如用户购买了某物品、用户查看了某物品的信息，等等。

显式的用户反馈能准确地反映用户对物品的真实喜好，但需要用户付出额外的代价；而隐式的用户行为通过一些分析和处理，也能反映用户的喜好，只是数据不是很精确。有些行为的分析存在较大的噪声，但只要选择正确的行为特征，隐式的用户反馈也能得到很好的效果，只是行为特征的选择可能在不同的应用中有很大的不同。例如，在电子商务的网站上，购买行为其实就是一个能很好表现用户喜好的隐式反馈。

推荐引擎根据不同的推荐机制可能用到数据源中的一部分，然后根据这些数据分析出一定的规则，或者直接对用户对其他物品的喜好进行预测计算。这样推荐引擎就可以在用户进入时给他推荐其可能感兴趣的物品。

ML 目前支持基于协同过滤的模型，在这个模型里，用户和产品被一组可以用来预测缺失项目的潜在因子描述，特别是我们实现交替最小二乘法来学习这些潜在的因子，ALS 算法原理如图 7-6 所示。

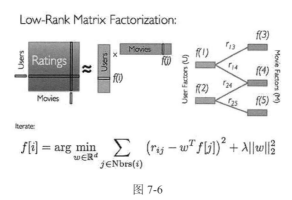

图 7-6

在 ML 中的实现有如下参数：

- numBlocks：用于并行化计算的分块个数（设置为-1 时为自动配置）。
- rank：模型中隐性因子的个数。
- iterations：迭代的次数。
- lambda：ALS 的正则化参数。
- implicitPrefs：决定是用显性反馈 ALS 的版本还是用隐性反馈 ALS 的版本。
- alpha：一个针对隐性反馈 ALS 版本的参数，这个参数决定了偏好行为强度的基准。

2. 案例说明

下面我们来看 Spark 3.0 版本中 ALS 算法的程序设计。

1）切分数据集

ALS 算法的前验基础是切分数据集，首先建立数据集文件 sample_movielens_ratings.txt，内容如图 7-7 所示。

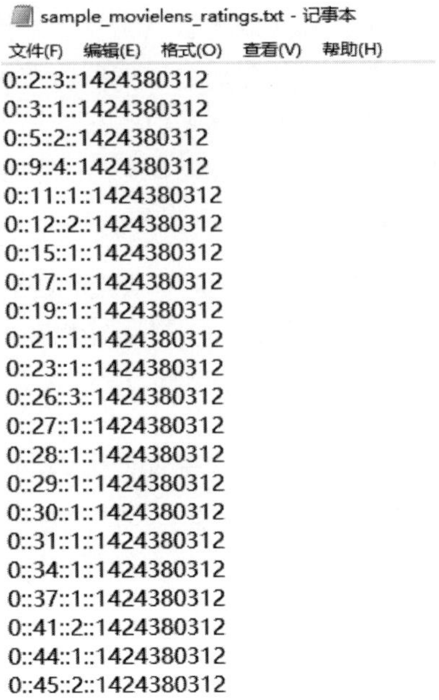
sample_movielens_ratings.txt - 记事本

文件(F) 编辑(E) 格式(O) 查看(V) 帮助(H)

0::2::3::1424380312
0::3::1::1424380312
0::5::2::1424380312
0::9::4::1424380312
0::11::1::1424380312
0::12::2::1424380312
0::15::1::1424380312
0::17::1::1424380312
0::19::1::1424380312
0::21::1::1424380312
0::23::1::1424380312
0::26::3::1424380312
0::27::1::1424380312
0::28::1::1424380312
0::29::1::1424380312
0::30::1::1424380312
0::31::1::1424380312
0::34::1::1424380312
0::37::1::1424380312
0::41::2::1424380312
0::44::1::1424380312
0::45::2::1424380312

图 7-7

需要注意的是，ML 中的 ALS 算法有固定的数据格式，源码如下：

```
case class Rating(userId: Int, movieId: Int, rating: Float, timestamp: Long)
```

其中，Rating 是固定的 ALS 输入格式，要求是一个元组类型的数据，其中的数值分别为[Int,Int, Float, Long]，因此建立在数据集时用户名和物品名分别用数值代替，而最终的评分没有变化，最后是一个时间戳（类型是 Long）。基于 Spark 3.0 架构，我们可以将迭代算法 ALS 很好地并行化。

2）建立 ALS 数据模型

ALS 数据模型是根据数据集训练获得的，源码如下：

```
val Array(training, test) = ratings.randomSplit(Array(0.8, 0.2))
val als = new ALS()
  .setMaxIter(5)
  .setRegParam(0.01)
  .setUserCol("userId")
  .setItemCol("movieId")
  .setRatingCol("rating")
val model = als.fit(training)
```

ALS 是由若干个 setters 设置参数构成的，其解释如下：

- numBlocks(numItemBlocks、numUserBlocks)：并行计算的 block 数（为-1 时表示自动配置）。
- rank：模型中的隐藏因子数。
- maxIter：算法最大迭代次数。
- regParam：ALS 中的正则化参数。

- implicitPref：使用显式反馈 ALS 变量或隐式反馈。
- alpha：ALS 隐式反馈变化率，用于控制每次拟合修正的幅度。
- coldStartStrategy：将 coldStartStrategy 参数设置为"drop"，以便删除 DataFrame 中包含 NaN 值的预测中的任何行。

这些参数协同作用，从而控制 ALS 算法的模型训练。

最终，Spark 3.0 ML 库基于 ALS 算法的协同过滤推荐代码如代码 7-4 所示。

代码 7-4 MovieLensALS.scala

```scala
import org.apache.spark.ml.evaluation.RegressionEvaluator
import org.apache.spark.ml.recommendation.ALS
import org.apache.spark.sql.SparkSession
object ALSExample {
  //定义 Rating 格式
  case class Rating(userId: Int, movieId: Int, rating: Float, timestamp: Long)
  def parseRating(str: String): Rating = {
    val fields = str.split("::")//分隔符
    assert(fields.size == 4)
    Rating(fields(0).toInt, fields(1).toInt, fields(2).toFloat, fields(3).toLong)
  }
  def main(args: Array[String]): Unit = {
    val spark = SparkSession
      .builder                        //创建 Spark 会话
      .master("local")                //设置本地模式
      .appName("ALSExample")          //设置名称
      .getOrCreate()                  //创建会话变量
    import spark.implicits._
    //读取 Rating 格式并转换 DF
    val ratings = spark.read.textFile("data/als/sample_movielens_ratings.txt")
      .map(parseRating)
      .toDF()
    val Array(training, test) = ratings.randomSplit(Array(0.8, 0.2))
    //在训练集上构建推荐系统模型、ALS 算法，并设置各种参数
    val als = new ALS()
      .setMaxIter(5)
      .setRegParam(0.01)
      .setUserCol("userId")
      .setItemCol("movieId")
      .setRatingCol("rating")
    val model = als.fit(training)//得到一个 model：一个 Transformer
    //在测试集上评估模型，标准为 RMSE
    //设置冷启动的策略为'drop'，以保证不会得到一个'NaN'的预测结果
    model.setColdStartStrategy("drop")
    val predictions = model.transform(test)
```

```
    val evaluator = new RegressionEvaluator()
      .setMetricName("rmse")
      .setLabelCol("rating")
      .setPredictionCol("prediction")
    val rmse = evaluator.evaluate(predictions)
    println(s"Root-mean-square error = $rmse")
    //为每一个用户推荐10部电影
    val userRecs = model.recommendForAllUsers(10)
    //为每部电影推荐10个用户
    val movieRecs = model.recommendForAllItems(10)
    //为指定的一组用户生成top10个电影推荐
    val users = ratings.select(als.getUserCol).distinct().limit(3)
    val userSubsetRecs = model.recommendForUserSubset(users, 10)
    //为指定的一组电影生成top10个用户推荐
    val movies = ratings.select(als.getItemCol).distinct().limit(3)
    val movieSubSetRecs = model.recommendForItemSubset(movies, 10)
    //打印结果
    userRecs.show()
    movieRecs.show()
    userSubsetRecs.show()
    movieSubSetRecs.show()
    spark.stop()
  }
}
```

在上面的程序中，使用 ALS()根据已有的数据集建立了一个协同过滤矩阵推荐模型，之后使用 recommendForAllUsers 方法为一个用户推荐 10 个物品（电影）等，结果打印如下：

```
Root-mean-square error = 1.684832316936912
+------+--------------------+
|userId|     recommendations|
+------+--------------------+
|    28|[[25, 6.00149], [...|
|    26|[[94, 5.29422], [...|
|    27|[[47, 6.3299623],...|
|    12|[[46, 6.5864477],...|
|    22|[[7, 5.437798], [...|
|     1|[[68, 3.8732295],...|
|    13|[[96, 3.8646204],...|
|     6|[[25, 4.5257554],...|
|    16|[[85, 4.960823], ...|
|     3|[[96, 4.1602864],...|
|    20|[[22, 4.770223], ...|
|     5|[[55, 4.090011], ...|
|    19|[[46, 5.232961], ...|
|    15|[[46, 4.8397903],...|
|    17|[[90, 4.914645], ...|
```

```
|     9|[[48, 5.1486597],...|
|     4|[[52, 4.2062426],...|
|     8|[[29, 5.071128], ...|
|    23|[[90, 5.842731], ...|
|     7|[[27, 5.47984], [...|
+------+--------------------+
only showing top 20 rows

+-------+--------------------+
|movieId|     recommendations|
+-------+--------------------+
|     31|[[12, 3.931116], ...|
|     85|[[16, 4.960823], ...|
|     65|[[23, 4.9316106],...|
|     53|[[21, 5.080318], ...|
|     78|[[0, 1.5588677], ...|
|     34|[[18, 4.6249347],...|
|     81|[[28, 4.7397876],...|
|     28|[[24, 5.2909055],...|
|     76|[[0, 4.9046974], ...|
|     26|[[11, 4.4119563],...|
|     27|[[7, 5.47984], [1...|
|     44|[[24, 4.7212014],...|
|     12|[[28, 4.688432], ...|
|     91|[[11, 3.1263103],...|
|     22|[[26, 5.134186], ...|
|     93|[[2, 5.194844], [...|
|     47|[[27, 6.3299623],...|
|      1|[[25, 2.9610748],...|
|     52|[[14, 4.997468], ...|
|     13|[[23, 3.8639143],...|
+-------+--------------------+
only showing top 20 rows

+------+--------------------+
|userId|     recommendations|
+------+--------------------+
|    28|[[25, 6.00149], [...|
|    26|[[94, 5.29422], [...|
|    27|[[47, 6.3299623],...|
+------+--------------------+

+-------+--------------------+
|movieId|     recommendations|
+-------+--------------------+
|     31|[[12, 3.931116], ...|
```

```
    |     85|[[16, 4.960823], ...|
    |     65|[[23, 4.9316106],...|
    +-------+--------------------+
```

在使用 ALS 进行预测时，通常会遇到测试数据集中的用户或物品没有出现的情况，这些用户或物品在训练模型期间不存在。针对上述问题，Spark 提供了将 coldStartStrategy 参数设置为 "drop" 的方式，就是删除 DataFrame 中包含 NaN 值的预测中的任何行，然后根据非 NaN 数据对模型进行评估，并且该评估是有效的。目前支持的冷启动策略是 "nan"（coldStartStrategy 参数的默认值）和 "drop"，将来可能会支持进一步的策略。

提示：程序中的 rank 表示隐藏因子，numIterator 表示循环迭代的次数，读者可以根据需要调节数值。报出 StackOverFlow 错误时，可以适当地调节虚拟机或者 IDE 的栈内存。另外，读者可以尝试调用 ALS 中的其他方法，以便更好地理解 ALS 模型的用法。Spark 官方实现的 ALS 算法由于调度方面的问题在训练的时候比较慢。

7.4.4 分类算法实战

1. 算法说明

分类算法是机器学习的重点。分类算法属于监督式学习，使用类标签为已知的样本建立一个分类函数或分类模型，并应用分类模型把数据集中的类标签未知的数据进行归类。分类在数据挖掘和机器学习中是一项重要的任务，目前在商业上应用最多，常见的典型应用场景有流失预测、精确营销、客户获取、个性偏好等。Spark ML 库目前支持的分类算法有：逻辑回归、支持向量机、朴素贝叶斯和决策树。

本节我们主要讲解比较典型的分类算法——朴素贝叶斯分类算法。

朴素贝叶斯分类器是机器学习中经典的分类模型，特点是易于理解且执行速度快，在针对多分类问题时其复杂度也不会有很大的上升。贝叶斯分类是第一种基于概率的分类方法，因贝叶斯公式而得名。朴素贝叶斯分类器是基于贝叶斯概率公式的一个朴素而有深度的模型。它的应用前提是样本特征之间相互独立，然后基于这些样本特征的条件概率乘积来计算每个分类的概率，最后选择概率最大的那个类别作为分类结果。

朴素贝叶斯分类算法是一种在分类领域易于理解且效果不错的分类方法，它以贝叶斯定理为基础，利用贝叶斯概率公式的特性将先验概率和条件概率转化为所求的后验概率，前提条件就是基于特征之间相互独立的假设。

朴素贝叶斯分类过程如下：

步骤01 假设 $x=\{a_1, a_2, a_3, \cdots, a_m\}$ 为一个样本，a_i 为 x 的一个特征属性。

步骤02 数据类别集合 $C=\{y_1, y_2, \cdots, y_n\}$。

步骤03 计算条件概率为 $P(y_1|x)$，$P(y_2|x)$，$P(y_3|x)$，$\cdots$，$P(y_n|x)$。

步骤04 经过计算假设 $P(y_k|x)=\max\{P(y_1|x), P(y_2|x), P(y_3|x), \cdots, P(y_n|x)\}$，那么样本 x 就划归到类别 y_k。

其中最关键的就是**步骤03** 条件概率的求解，此处涉及的就是贝叶斯概率公式：首先要计算每个

类别的样本数据集大小，确定该类别下每个特征属性的条件概率，即第 i 个类别对应的第 j 个特征下的条件概率为 $P(a_j|y_i)$，其中 $i=1,2,\cdots,n$，$j=1,2,\cdots,m$。然后基于特征之间条件独立的假设，依据贝叶斯定理最终推导出计算公式为：

$$p(y_i|x) = \frac{p(x|y_i)p(y_i)}{p(x)}$$

下面通过一个例子来帮助读者理解朴素贝叶斯分类算法。

假设某个医院早上已收治了 6 个门诊病人，如下：

症状	职业	疾病
打喷嚏	护士	感冒
打喷嚏	农夫	过敏
头痛	建筑工人	脑震荡
头痛	建筑工人	感冒
打喷嚏	教师	感冒
头痛	教师	脑震荡

现在又来了第 7 个病人，是一个打喷嚏的建筑工人。请问他患上感冒的概率有多大？

根据贝叶斯定理 $P(A|B) = P(B|A) P(A) / P(B)$ 可得：

$P(感冒|打喷嚏×建筑工人)= P(打喷嚏×建筑工人|感冒)×P(感冒)/P(打喷嚏×建筑工人)$

假定"打喷嚏"和"建筑工人"这两个特征是独立的，因此，上面的等式就变成：

$P(感冒|打喷嚏×建筑工人)=P(打喷嚏|感冒)×P(建筑工人|感冒)×P(感冒)/P(打喷嚏)×P(建筑工人)$

这是可以计算的，算式如下：

$P(感冒|打喷嚏×建筑工人)= 0.66×0.33×0.5 / 0.5×0.33= 0.66$

因此，这个打喷嚏的建筑工人，有 66%的概率是得了感冒。同理，可以计算这个病人患上过敏或脑震荡的概率。比较这几个概率，就可以知道他最可能得什么病。

上述就是贝叶斯分类算法：在统计资料的基础上，依据某些特征计算各个类别的概率，从而实现分类。

2. 案例实战

本案例主要基于鸢尾花数据集进行分类，首先对数据集进行说明。

Iris 鸢尾花数据集包含 3 类鸢尾花，分别为山鸢尾（Iris-setosa）、变色鸢尾（Iris-versicolor）和维吉尼亚鸢尾（Iris-virginica），如图 7-8 所示。数据集共 150 条数据，每类各 50 条数据，每条数据都有 4 项特征：花萼长度、花萼宽度、花瓣长度、花瓣宽度。通常可以通过这 4 项特征预测鸢尾花卉属于哪一品种。即数据集中有 4 类观测特征和一个判定归属，一共有 150 条数据。更进一步说，每条数据的记录是观测一个鸢尾花瓣所具有的不同特征数，即：

- 萼片长（sepal length）
- 萼片宽（sepal width）

- 花瓣长（petal length）
- 花瓣宽（petal width）
- 种类（species)

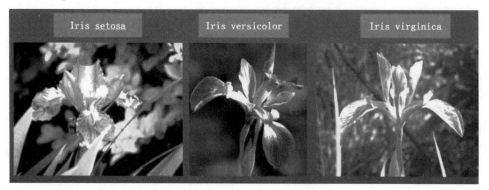

图 7-8

本算法采用的数据集 iris.data 文件格式如图 7-9 所示。

```
5.1,3.5,1.4,0.2,Iris-setosa
4.9,3.0,1.4,0.2,Iris-setosa
4.7,3.2,1.3,0.2,Iris-setosa
4.6,3.1,1.5,0.2,Iris-setosa
5.0,3.6,1.4,0.2,Iris-setosa
5.4,3.9,1.7,0.4,Iris-setosa
4.6,3.4,1.4,0.3,Iris-setosa
5.0,3.4,1.5,0.2,Iris-setosa
4.4,2.9,1.4,0.2,Iris-setosa
4.9,3.1,1.5,0.1,Iris-setosa
5.4,3.7,1.5,0.2,Iris-setosa
4.8,3.4,1.6,0.2,Iris-setosa
4.8,3.0,1.4,0.1,Iris-setosa
```

图 7-9

ML 库中贝叶斯方法主要是作为多类分类器来使用的，是一系列基于朴素贝叶斯的算法。基于贝叶斯定理，每对特征之间具有强（朴素）独立性假设，即所有朴素贝叶斯分类器都假定样本的每个特征与其他特征都不相关，其目的是根据向量的不同对它进行分类处理。

朴素贝叶斯可以非常有效地用于训练。通过对训练数据的单次传递，它计算每个给定标签的每个特征的条件概率分布。对于预测，它应用贝叶斯定理计算给定观测值的每个标签的条件概率分布。

本案例涉及多分类，需要采用 MulticlassClassificationEvaluator 来实现。完整的鸢尾花贝叶斯分类实现代码如下：

代码 7-5 naive_bayes.scala

```
import org.apache.spark.SparkConf
import org.apache.spark.ml.classification.NaiveBayes
import org.apache.spark.ml.evaluation.MulticlassClassificationEvaluator
```

```scala
import org.apache.spark.ml.feature.VectorAssembler
import org.apache.spark.sql.SparkSession
import scala.util.Random
/**
 * Author : mrchi
 * Time   : 2023/1/9
 **/
object naive_bayes extends App {
  val conf = new SparkConf().setMaster("local").setAppName("iris")
  val spark = SparkSession.builder().config(conf).getOrCreate()
  spark.sparkContext.setLogLevel("WARN")  ///日志级别
  val file = spark.read.format("csv").load("iris.data")
  import spark.implicits._
  val random = new Random()
  val data = file.map(row => {
    val label = row.getString(4) match {
      case "Iris-setosa" => 0
      case "Iris-versicolor" => 1
      case "Iris-virginica" => 2
    }
    (row.getString(0).toDouble,
      row.getString(1).toDouble,
      row.getString(2).toDouble,
      row.getString(3).toDouble,
      label,
      random.nextDouble())
  }).toDF("_c0", "_c1", "_c2", "_c3", "label", "rand").sort("rand")
  //  data.show()
  val assembler = new VectorAssembler().setInputCols(Array("_c0", "_c1", "_c2", "_c3")).setOutputCol("features")
  val dataset = assembler.transform(data)
  val Array(train, test) = dataset.randomSplit(Array(0.8, 0.2))
  //bayes
  val bayes = new NaiveBayes().setFeaturesCol("features").setLabelCol("label")
  val model = bayes.fit(train)  //训练数据集进行训练
  val result = model.transform(test) //测试数据集进行测试,看看效果如何
  val evaluator = new MulticlassClassificationEvaluator()
    .setLabelCol("label")
    .setPredictionCol("prediction")
    .setMetricName("accuracy")
  val accuracy = evaluator.evaluate(result)
  println(s"""accuracy is $accuracy""")
}
```

需要说明的是，从 Spark 3.0 开始，ML 库开始支持 Complement Naive Bayes（补充朴素贝叶斯，是标准多项式朴素贝叶斯的改进形式）以及高斯朴素贝叶斯（Gaussian Naive Bayes，可以处理连续数据）。Multinomial Naive Bayes、Bernoulli Naive Bayes、Complement Naive Bayes 通常用于文档分类，使用可选参数"Multinomial""Complement""Bernoulli"或"Gaussian"选择模型类型，默认为"Multinomial"。Spark 还提供了一种叫作平滑操作的技术，对于测试集中的一个类别变量特征，如果在训练集里没有见过，那么直接计算的话其概率就是 0，而平滑操作可以缓解预测功能失效的问题。

第 8 章

Spark GraphX 图计算

GraphX 是 Spark 用于图和分布式图（graph-parallel）计算的新组件。GraphX 通过引入弹性分布式属性图——顶点和边均有属性的有向多重图，来扩充 Spark RDD。为了支持这种图计算，目前 GraphX 开发了一组基础功能操作。GraphX 仍在不断扩充图算法，以简化图计算的分析任务。

本章主要知识点：

- Spark GraphX 概述
- Spark GraphX 的抽象
- Spark GraphX 图的构建
- Spark GraphX 图的计算模式
- Spark GraphX 实战

8.1 Spark GraphX

GraphX 是一个新的 Spark API，它用于图和分布式图的计算。GraphX 通过引入弹性分布式属性图（Resilient Distributed Property Graph）——顶点和边均有属性的有向多重图，来扩展 Spark RDD。为了支持图计算，GraphX 公开了一系列基本运算符（比如 mapVertices、mapEdges、subgraph）以及优化后的 Pregel API 变种。此外，还包含越来越多的图算法和构建器，以简化图形分析任务。

Spark GraphX 是 Spark 的一个模块，主要用于进行以图为核心的计算以及分布式图的计算。GraphX 的底层计算也是 RDD 计算，它和 RDD 共用一种存储形态，在展示形态上可以用数据集来表示，也是以图的形式来表示。

图数据很好地表达了数据之间的相关性，现实中很多问题都可以用图来表示。以下的一些问题就可以用图的数据结构模型来解决问题。

- PangeRank 让链接来投票。

- 基于 GraphX 的社区发现算法。
- 社交网络分析。
- 基于三角形计数关系的衡量。
- 基于随机游走的用户属性传播。
- 推荐应用。

GraphX 的整体架构可以分为 3 层：

- 算法层：基于 Pregel 接口实现了常用的图算法。包括 PageRank、SVDPlusPlus、TriangleCount、ConnectedComponents、StronglyConnectedConponents 等算法。
- 接口层：在底层 RDD 的基础之上实现了 Pregel 模型、BSP 模型的计算接口。
- 底层：图计算的核心类，包含 VertexRDD、EdgeRDD、RDD[EdgeTriplet]。

8.2　Spark GraphX 的抽象

1. 顶点

顶点用 RDD[(VertexId, VD)]来表示，其中 VertexID 表示顶点的 ID，是 Long 类型的别名；VD 是顶点的属性，是一个类型参数，可以是任何类型。

2. 边

边用 RDD[Edge[ED]]来表示，Edge 用来具体表示一个边，Edge 里面包含一个用 ED 类型参数来设定的属性、一个源顶点的 ID 和一个目标顶点的 ID。

3. 三元组

三元组结构用 RDD[EdgeTriplet[VD, ED]]来表示，三元组包含了一条边、边的属性、源顶点 ID、源顶点属性、目标顶点 ID、目标顶点属性。VD 和 ED 是类型参数，VD 表示顶点的属性，ED 表示边的属性。

通过 Triplet 成员，用户可以直接获取起点顶点、起点顶点属性、终点顶点、终点顶点属性、边与边属性等信息。Triplet 的生成可以由边表与顶点表通过 ScrId 与 DstId 连接而成。Triplet 对应着 EdgeTriplet，它是一个三元组视图，这个视图在逻辑上将顶点和边的属性保存为一个 RDD[EdgeTriplet[VD, ED]]。

4. 图

图在 Spark 中用 Graph 来表示，可以通过顶点和边来构建。

GraphX 用属性图的方式表示图，顶点有属性，边有属性。存储结构采用边集数组的形式，即一个顶点表，一个边表，如图 8-1 所示。

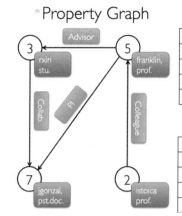

图 8-1

顶点 ID 是非常重要的字段，它不光是顶点的唯一标识符，也是描述边的唯一手段。顶点表与边表实际上就是 RDD，它们分别为 VertexRDD 与 EdgeRDD。

- vertices 为顶点表，VD 为顶点属性类型。
- edges 为边表，ED 为边属性类型。
- 可以通过 Graph 的 vertices 与 edges 成员直接得到顶点 RDD 与边 RDD。
- 顶点 RDD 类型为 VerticeRDD，继承自 RDD[(VertexId, VD)]。
- 边 RDD 类型为 EdgeRDD，继承自 RDD[Edge[ED]]。

8.3　Spark GraphX 图的构建

通常在图计算中，基本的数据结构表达就是 G=（V，E，D），其中，V 为 vertex（顶点或者节点），E 为 edge（边），D 为 data（权重）。

1. 顶点的构建

对于 RDD[(VertexId, VD)]这种版本：

```
val users: RDD[(VertexId, (String, String))] = sc.parallelize(Array((3L, ("rxin", "student")), (7L, ("jgonzal", "postdoc")),(5L, ("franklin", "prof")), (2L, ("istoica", "prof"))))
```

对于 VertexRDD[VD]这种版本（是顶点的优化版本）：

```
val users1:VertexRDD[(String, String)] = VertexRDD(String, String)
```

2. 边的构建

对于 RDD[Edge[ED]]这种版本：

```
val relationships: RDD[Edge[String]] = sc.parallelize(Array(Edge(3L, 7L, "collab"), Edge(5L, 3L, "advisor"),Edge(2L, 5L, "colleague"), Edge(5L, 7L, "pi")))
```

对于 EdgeRDD[ED]这种版本（是边的优化版本）：

```
val relationships1:EdgeRDD[String] = EdgeRDD.fromEdges(relationships)
```

3. Graph 图的构建

通过 Graph 类的 apply 方法进行构建：Graph[VD: ClassTag, ED: ClassTag]

```
val graph = Graph(users,relationships)
```

apply 方法：

```
def apply[VD: ClassTag, ED: ClassTag](
   vertices: RDD[(VertexId, VD)],
   edges: RDD[Edge[ED]],
   defaultVertexAttr: VD = null.asInstanceOf[VD],
   edgeStorageLevel: StorageLevel = StorageLevel.MEMORY_ONLY,
   vertexStorageLevel: StorageLevel = StorageLevel.MEMORY_ONLY): Graph[VD, ED]
```

通过 Graph 类提供的 fromEdges 方法来构建，顶点的属性使用提供的默认属性。

```
val graph2 = Graph.fromEdges(relationships,defaultUser)
def fromEdges[VD: ClassTag, ED: ClassTag](
   edges: RDD[Edge[ED]],
   defaultValue: VD,
   edgeStorageLevel: StorageLevel = StorageLevel.MEMORY_ONLY,
   vertexStorageLevel: StorageLevel = StorageLevel.MEMORY_ONLY): Graph[VD, ED]
```

通过 Graph 类提供的 fromEdgeTuples 方法来构建，顶点的属性使用提供的默认属性，边的属性是使用相同边的数量。

```
    val relationships: RDD[(VertexId,VertexId)] = sc.parallelize(Array((3L,
7L),(5L, 3L),(2L, 5L), (5L, 7L)))
    val graph3 = Graph.fromEdgeTuples[(String,String)](relationships,
defaultUser)
   def fromEdgeTuples[VD: ClassTag](
      rawEdges: RDD[(VertexId, VertexId)],
      defaultValue: VD,
      uniqueEdges: Option[PartitionStrategy] = None,
      edgeStorageLevel: StorageLevel = StorageLevel.MEMORY_ONLY,
      vertexStorageLevel: StorageLevel = StorageLevel.MEMORY_ONLY): Graph[VD,
Int]
```

对于图 8-1 所示的属性图，Graph 图的构建代码如下：

代码 8-1　CreateGraph .scala

```
object CreateGraph {
  def main(args: Array[String]): Unit = {
    // 创建 SparkContext 对象
    val sc: SparkContext = SparkContext.getOrCreate(new
SparkConf().setMaster("local[*]").setAppName("CreateGraph"))
    // 创建保存顶点信息的 RDD
    val users: RDD[(VertexId, (String, String))] =
```

```
    sc.parallelize(Array((3L, ("rxin", "student")), (7L, ("jgonzal",
"postdoc")),
      (5L, ("franklin", "prof")), (2L, ("istoica", "prof")))))
    // 创建保存边信息的 RDD
    val relationships: RDD[Edge[String]] =
      sc.parallelize(Array(Edge(3L, 7L, "collab"), Edge(5L, 3L, "advisor"),
        Edge(2L, 5L, "colleague"), Edge(5L, 7L, "pi")))
    // 定义一个默认用户
    val defaultUser = ("John Doe", "Missing")
    // 构建图对象
    val graph = Graph(users, relationships, defaultUser)
    // 测试图对象
    graph.triplets.foreach(triple =>
println(s"(${triple.srcId},${triple.srcAttr}) =(${triple.attr})=>
(${triple.dstId},${triple.dstAttr})"))
  }
}
```

8.4　Spark GraphX 图的计算模式

1. Spark 视图

Spark 中有 3 种视图，即 vertices、edges 和 triplets，这 3 种视图可以分别通过 graph.vertices、graph.edges 和 graph.triplets 来访问。

1）通过 case 访问

用 case 语句进行形式简单、功能强大的模式匹配，示例如下：

```
val graph: Graph[(String, Int), Int] = Graph(vertexRDD, edgeRDD)
```

用 case 匹配可以很方便访问顶点和边的属性及 ID：

```
graph.vertices.map{
    case (id,(name,age))=>//利用 case 进行匹配
      (age,name)//可以在这里加上自己想要的任何转换
}
graph.edges.map{
    case Edge(srcid,dstid,weight)=>//利用 case 进行匹配
      (dstid,weight*0.01)//可以在这里加上自己想要的任何转换
}
```

2）通过下标访问

通过下标访问顶点和边的属性及 ID：

```
graph.vertices.map{
    v=>(v._1,v._2._1,v._2._2)//v._1、v._2._1、v._2._2 分别对应 ID、name、age
}
graph.edges.map {
    e=>(e.attr,e.srcId,e.dstId)
```

```
    }
    graph.triplets.map{
        triplet=>(triplet.srcAttr._1,triplet.dstAttr._2,triplet.srcId,
triplet.dstId)
    }
```

2. Spark GraphX 图的基本信息转换

- graph.numEdges：返回当前图的边的数量。
- graph.numVertices：返回当前图的顶点的数量。
- graph.inDegrees：返回当前图每个顶点入度的数量，返回类型为 VertexRDD[Int]。
- graph.outDegrees：返回当前图每个顶点出度的数量，返回的类型为 VertexRDD[Int]。
- graph.degrees：返回当前图每个顶点入度和出度的和，返回的类型为 VertexRDD[Int]。

3. Spark GraphX 图的转换操作

- def mapVertices[VD2: ClassTag](map: (VertexId, VD) => VD2) (implicit eq: VD =:= VD2 = null): Graph[VD2, ED]：对当前图的每一个顶点应用 map 函数来修改顶点的属性，返回一个新的图。
- def mapEdges[ED2: ClassTag](map: Edge[ED] => ED2): Graph[VD, ED2]：对当前图的每一条边应用 map 函数来修改边的属性，返回一个新图。
- def mapTriplets[ED2: ClassTag](map: EdgeTriplet[VD, ED] => ED2): Graph[VD, ED2]。对当前图的每一个三元组应用 map 函数来修改边的属性，返回一个新图。

4. Spark GraphX 图的结构操作

- def reverse: Graph[VD, ED]：该操作反转一个图，产生一个新图，新图中的每条边的方向和原图每条边的方向相反。
- def subgraph(epred: EdgeTriplet[VD, ED] => Boolean = (x => true), vpred: (VertexId, VD) => Boolean = ((v, d) => true)) : Graph[VD, ED]：该操作返回一个当前图的子图，通过传入 epred 函数来过滤边，通过传入 vpred 函数来过滤顶点，返回由满足 epred 函数值为 true 的边和满足 vpred 函数值为 true 的顶点组成子图。
- def mask[VD2: ClassTag, ED2: ClassTag](other: Graph[VD2, ED2]): Graph[VD, ED]：mask 函数用于求一个图和 other 这个图的交集，该交集的判别条件指的是：① 对于顶点，只对比顶点的 ID；② 对于边，只对比边的 srcID、dstID，如果 other 和当前图的交集中的边、顶点的属性不一致，那么 mask 产生的图默认采用当前图的属性。
- def groupEdges(merge: (ED, ED) => ED): Graph[VD, ED]：该操作实现将当前图中的两条相同边（边的 srcID 和 dstID 相同）合并。我们需要传入一个 merge 函数，用于合并这两边的属性并返回一个新的属性。

5. 关联操作

- joinVertices：将外部数据加入图中，操作 join 输入 RDD 和顶点。当图中的某个顶点 ID 在另一个图中不存在时，会保留原图中该顶点属性的原值。
- outerJoinVertices：将外部数据加入图中，操作 join 输入 RDD 和顶点。当图中的某个顶点 ID 在另一个图中不存在时，会使用 None 值作为该顶点的属性值。

6. 聚合操作

- aggregateMessages：主要功能是向邻边发消息，合并邻边收到的消息，返回 messageRDD。

例如，计算比用户年龄大的追随者（即 followers）的平均年龄，示例代码如下：

```
import org.apache.spark.graphx.util.GraphGenerators
val graph: Graph[Double, Int] = GraphGenerators.lognormalGraph(sc, numVertices = 100).mapvertices((id, _) => id.toDouble)
val olderFollowers: VertexRDD[(Int, Double)] = graph.aggregateMessages[(Int, Double)](
    triplet => {
        if (triplet.srcAttr > triplet.dstAttr){
            triplet.sendToDst(1, triplet.srcAttr)
        }
    },
    (a, b) => (a._1 + b._1, a._2 + b._2)
)
val avgageofolderFollowers: VertexRDD[Double] = olderFollowers.mapvalues((id, value) => value match{
    case (count, totalAge) => totalAge / count})
avgageofolderFollowers.collect.foreach(printIn(_))
```

8.5 GraphX 3 个主要算法实战

GraphX 中自带一系列图算法来简化分析，这些算法在 org.apache.spark.graphx.lib 包中，可以被 Graph 通过 GraphOps 直接访问。本节主要介绍 GraphX 中的 3 个主要算法。

1. PageRank 算法

PageRank 又称网页排名算法，它通过网络的超链接关系确定网页的等级，在搜索引擎优化操作中常用来评估网页的相关性和重要性。

PageRank 同样可以在图中测量每个顶点的重要性，假设存在一条从顶点 u 到顶点 v 的边，就代表顶点 u 对顶点 v 的支持。例如，在微博中一个用户被其他很多用户关注，那么这个用户的排名将会很高。GraphX 自带静态和动态的 PageRank 算法实现：静态的 PageRank 算法运行固定的迭代次数，动态的 PageRank 算法运行直到整个排名收敛（通过限定可容忍的值来停止迭代）。

本实战利用 GraphX 自带的社会网络数据集实例，用户集合数据集保存在/usr/local/Spark/data/graphx/users.txt 文件中，用户关系数据集保存在/usr/local/Spark/data/graphx/followers.txt 文件中。现在需要计算每个用户的 PageRank，代码如下：

代码 8-2 SimpleGraphX.scala

```
import org.apache.log4j.{Level,Logger}
import org.apache.spark._
import org.apache.spark.graphx.GraphLoader
```

```
object SimpleGraphX {
  def main(args: Array[String]) {
    //屏蔽日志
    Logger.getLogger("org.apache.spark").setLevel(Level.WARN)
    Logger.getLogger("org.eclipse.jetty.server").setLevel(Level.OFF)
    //设置运行环境
    val conf = new SparkConf().setAppName("SimpleGraphX").setMaster("local")
    val sc = new SparkContext(conf)
    // Load the edges as a graph
    val graph = GraphLoader.edgeListFile(sc, "followers.txt")
    // Run PageRank
    val ranks = graph.pageRank(0.0001).vertices
    // Join the ranks with the usernames
    val users = sc.textFile("users.txt").map { line =>
      val fields = line.split(",")
      (fields(0).toLong, fields(1))
    }
    val ranksByUsername = users.join(ranks).map {
      case (id, (username, rank)) => (username, rank)
    }
    // Print the result
    println(ranksByUsername.collect().mkString("\n"))
  }
}
```

2. 连通分支算法

连通分支算法使用最小编号的顶点来标记每个连通分支。在一个社会网络中，连通图近似簇。这里我们计算一个连通分支实例，所使用的数据集和 PageRank 一样。代码如下：

代码 8-3　SimpleGraphLiantong.scala

```
import org.apache.log4j.{Level,Logger}
import org.apache.spark._
import org.apache.spark.graphx.GraphLoader
object SimpleGraphXLiantong {
  def main(args: Array[String]) {
    //屏蔽日志
    Logger.getLogger("org.apache.spark").setLevel(Level.WARN)
    Logger.getLogger("org.eclipse.jetty.server").setLevel(Level.OFF)
    //设置运行环境
    val conf = new SparkConf().setAppName("SimpleGraphX").setMaster("local")
    val sc = new SparkContext(conf)
    // Load the edges as a graph
    // Load the graph as in the PageRank example
    val graph = GraphLoader.edgeListFile(sc, "followers.txt")
    // Find the connected components
    val cc = graph.connectedComponents().vertices
    // Join the connected components with the usernames
    val users = sc.textFile("users.txt").map { line =>
      val fields = line.split(",")
```

```
      (fields(0).toLong, fields(1))
    }
    val ccByUsername = users.join(cc).map {
      case (id, (username, cc)) => (username, cc)
    }
    // Print the result
    println(ccByUsername.collect().mkString("\n"))
  }
}
```

3. 三角形计算算法

在图中，如果一个顶点有两个邻接顶点并且顶点与顶点之间有边相连，那么我们就可以把这三个顶点归于一个三角形。

这里计算社交网络图中三角形的数量，所采用的数据集和 PageRank 的数据集一样。代码如下：

代码 8-4　SimpleGraphTriangle.scala

```
import org.apache.log4j.{Level,Logger}
import org.apache.spark._
import org.apache.spark.graphx.{GraphLoader,PartitionStrategy}
object SimpleGraphTriangle {
  def main(args: Array[String]) {
    //屏蔽日志
    Logger.getLogger("org.apache.spark").setLevel(Level.WARN)
    Logger.getLogger("org.eclipse.jetty.server").setLevel(Level.OFF)
    //设置运行环境
    val conf = new SparkConf().setAppName("SimpleGraphX").setMaster("local")
    val sc = new SparkContext(conf)
    // Load the edges in canonical order and partition the graph for triangle count
    val graph = GraphLoader.edgeListFile(sc, "followers.txt", true)
      .partitionBy(PartitionStrategy.RandomVertexCut)
    // Find the triangle count for each vertex
    val triCounts = graph.triangleCount().vertices
    // Join the triangle counts with the usernames
    val users = sc.textFile("users.txt").map { line =>
      val fields = line.split(",")
      (fields(0).toLong, fields(1))
    }
    val triCountByUsername = users.join(triCounts).map { case (id, (username, tc)) =>
      (username, tc)
    }
    // Print the result
    println(triCountByUsername.collect().mkString("\n"))
  }
}
```

读者可自行打印验证以上 3 个算法的结果。

8.6 GraphX 综合应用项目实战

在说明应用需求之前,我们需要提前为 GraphX 的应用项目做好准备工作。

首先,在本项目中引入 GraphX 的依赖,代码如下:

```xml
<dependency>
    <groupId>org.apache.spark</groupId>
    <artifactId>spark-graphx_2.12</artifactId>
    <version>3.0.0</version>
</dependency>
```

然后引入 Spark 和 GraphX 框架到项目中,代码如下:

```
import org.apache.spark._
import org.apache.spark.graphx._
// 为了使一些例子能正常工作,还需引入 RDD
import org.apache.spark.rdd.RDD
```

如果没有用到 Spark shell,那么还将需要 SparkContext。

具体需求为:定义 Graph 中有 6 个人,每个人有名字和年龄,这些人根据社会关系形成 8 条边,每条边都有属性,如图 8-2 所示。本综合应用实战将构建顶点、边和图,打印图的属性,进行转换操作、结构操作、连接操作、聚合操作,并结合实际要求进行演示,具体代码及注释如代码 8-5 所示。

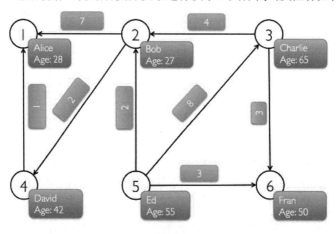

图 8-2

代码 8-5 GraphXExample.scala

```scala
import org.apache.log4j.{level, Logger}
import org.apache.spark.{Sparkcontext, Sparkconf}
import org.apache.spark.graphx._
import org.apache.spark.rdd.RDD
object GraphXExample {
    def main(args: Array[String]) {
//屏蔽日志
Logger.getLogge("org.apache.spark").setLevel(Level.WARN)
```

```scala
//设置运行环境
val conf = new Sparkconf().setAppName("SimpleGraphX").setMaster("local")
val sc = new Sparkcontext(conf)
//设置顶点和边，注意顶点和边都是用元组定义的 Array
//顶点的数据类型是 VD:(String,int)
val vertexArray = Array(
    (1L, ("Alice", 28)),
    (2L, ("Bob", 27)),
    (3l, ("Charlie", 65)),
    (4l, ("David", 42)),
    (5L, ("Ed", 55)),
    (6l, ("Fran", 50))
)
//边的数据类型 ED:int
val edgeArray = Array(
    Edge(2L, 1L, 7),
    Edge(2L, 4L, 2),
    Edge(3L, 2L, 4),
    Edge(3L, 6L, 3),
    Edge(4L, 1L, 1),
    Edge(5L, 2L, 2),
    Edge(5L, 3L, 8),
    Edge(5L, 6L, 3)
)
//构造 vertexRDD 和 edgeRDD
val vertexRDD: RDD[(Long, (String, int))] = sc.paral1elize(vertexArray)
val edgeRDD: RDD[Edge[Int]] = sc.paral1elize(edgeArray)
//构造图 Graph[VD, ED]
val graph: Graph[(String, Int), Int] = Graph(vertexRDD, edgeRDD)
printIn("属性演示")
printIn("***********************")
printIn("找出图中年龄大于 30 的顶点：")
graph.vertices.filter{
    case(id, (name, age)) => age >30}.collect.foreach{
        case(id, (name, age)) => printIn(s"$name is $age")
    }
printIn("找出图中属性大于 5 的边：")
graph.edges.filter(e => e.attr > 5).collect.foreach(e =>printIn(s"${e.srcId} to ${e.dstId} att ${e.attr}"))
//triplets 操作，((srcid, srcAttr), (dstid, dstAttr), attr)
printIn("列出边属性>5 的tripltes:")
for (triplet <- graph.triplets.filter(t => t.attr > 5).collect) {
    println(s"${triplet.srcAttr._1} likes ${triplet.dstAttr._1}")
}
println
//Degrees 操作
```

```
println("找出图中最大的出度、入度度数：")
def max(a: (vertexid, int), b: (vertexid, int)): (vertexid, int) = {
    if (a._2 > b._2) a else b
}
println("max of outDegrees:" + graph.outDegrees.reduce(max) + " max of inDegrees:" + graph.inDegrees.reduce(max) + " max of Degrees:" + graph.degrees.reduce(max))
println
//**********转换操作***********
prIntln("顶点的转换操作，顶点age + 10:")
graph.mapVertices{ case (id, (name, age)) => (id, (name, age+10))}.vertices.collect.foreach(v => println(s"${v._2._1} is ${v._2._2}"))
println
pHntln("边的转换操作，边的属性*2:")
graph.mapEdges(e=>e.attr*2).edges.collect.foreach(e => prIntln(s"${e.srcid} to ${e.dstid} att ${e.attr}"))
println
//******结构操作********
prIntln("结构操作")
pHntln("顶点年纪>30 的子图：")
val subGraph = graph.subgraph(vpred = (id, vd) => vd._2 >= 30)
println("子图所有顶点：")
subGraph.vertices.collect.foreach(v => println(s"${v._2._1} is ${v__2._2}"))
prIntln
pHntln("子图所有边：")
subGraph.edges.collect.foreach( e => prIntln(s"${e.srcId} to ${e.dstId} att ${e.attr}"))
println
//**********连接操作***************
prIntln("连接操作")
val inDegrees: VertexRDD[Int] = graph.inDegrees
case class User(name: String, age: Int, inDeg:Int, outDeg: Int)
//创建一个新图，顶点VD的数据类型为User，并从Graph做类型转换
val initialUserGraph: Graph[User, Int] = graph.mapVertices{
    case(id, (name, age)) => User(name, age, 0, 0)}
    val userGraph = initialUserGraph.outerJoinVertices( initialUserGraph.inDegrees) {
    case (id, u, inDegOpt) => user(u.name, u.age, inDegOpt.getOrElse(0), u.outDeg)
}.outerJoinVertices(initialUserGraph.outDegrees) {
    case (id, u, outDegOpt) => user(u.name, u.age,
    u.inDeg,outDegopt.getOrElse(0))
}
println("连接图的属性：")
userGraph.vertices.collect.foreach(v => println(s"${v._2.name} inDeg:
```

```
${v._2.inDeg} outDeg: ${v._2.outDeg}"))
println
println("出度和入度相同的人员：")
userGraph.vertices.filter {
case (id, u) => u.inDeg == u.outDeg
}.collect.foreach {
case (id, property) => println(property.name)
println
//**********聚合操作**********
println("找出年纪最大的追求者：")
val oldestFollower: vertexRDD[(String, Int)]=
userGraph.mapReduceTriplets[(String, Int)](
//将源顶点的属性发送给目标顶点，map 过程
edge => iterator((edge.dstid, (edge.srcAttr.name, edge.srcAttr.age))),
//得到最大追求者，reduce 过程
(a, b) => if (a._2 > b._2) a else b
)
userGraph.vertices.leftJoin(oldestFollower) { (id, user, optOldestFollower) =>
optoldestFollower match {
    case None => s"${user.name} does not have any followers."
    case some((name, age)) => s"${name} is the oldest follower of
${user.name}."
}
}.collect.foreach { case (id, str) => println(str)}
println
//***********聚合操作***********
println("找出 5 到各顶点的最短距离：")
val sourceId: vertexId = 5L  // 定义源点
val initialGraph = graph.mapVertices((id, _) => if (id == sourceId) 0.0
else Double.Positivelnfinity)
val sssp = initialGraph.pregel(Double.Positivelnfinity)(
(id, dist, newDist) => math.min(dist, newDist),
triplet => {  // 计算权重
if (triplet.srcAttr + triplet.attr < triplet.dstAttr) {
    Iterator((triplet.dstld, triplet.srcAttr + triplet.attr))
} else {
    Iterator.empty
}
},
(a,b) => math, min (a, b) //最短距离
)
println(sssp.vertices.collect.mkstring("/n"))
sc.stop()
    }
}
```

首先对 GraphXExample.scala 代码进行编译，编译通过后开始执行，执行结果如下：

```
************************************************************
属性演示
************************************************************

找出图中年龄大于 30 的顶点:

David is 42

Fran is 50

Charlie is 65

Ed is 55

找出图中属性大于 5 的边:

2 to 1 att 7

5 to 3 att 8

列出边属性>5 的 tripltes:

Bob likes Alice

Ed likes Charlie

找出图中最大的出度数和入度数:

max of outDegrees:(5,3) max of inDegrees:(2,2) max of Degrees:(2,4)

************************************************************
转换操作
************************************************************

顶点的转换操作, 顶点 age + 10:

4 is (David,52)

1 is (Alice,38)

6 is (Fran,60)

3 is (Charlie,75)

5 is (Ed,65)
```

2 is (Bob,37)

边的转换操作，边的属性*2：

2 to 1 att 14

2 to 4 att 4

3 to 2 att 8

3 to 6 att 6

4 to 1 att 2

5 to 2 att 4

5 to 3 att 16

5 to 6 att 6

****
结构操作
****

顶点年纪>30 的子图：

子图所有顶点：

David is 42

Fran is 50

Charlie is 65

Ed is 55

子图所有边：

3 to 6 att 3

5 to 3 att 8

5 to 6 att 3

****
连接操作

```
************************************************************
```

连接图的属性：

David inDeg: 1 outDeg: 1

Alice inDeg: 2 outDeg: 0

Fran inDeg: 2 outDeg: 0

Charlie inDeg: 1 outDeg: 2

Ed inDeg: 0 outDeg: 3

Bob inDeg: 2 outDeg: 2

出度和入度相同的人员：

David

Bob

```
************************************************************
```
聚合操作
```
************************************************************
```

找出年纪最大的追求者：

Bob is the oldest follower of David.

David is the oldest follower of Alice.

Charlie is the oldest follower of Fran.

Ed is the oldest follower of Charlie.

Ed does not have any followers.

Charlie is the oldest follower of Bob.

```
************************************************************
```
实用操作
```
************************************************************
```

找出 5 到各顶点的最短距离：

(4,4.0)

(1,5.0)

(6,3.0)

(3,8.0)

(5,0.0)

(2,2.0)

第 9 章

Redis 数据库入门

Redis 是一款开源的、基于 BSD 许可的高级键值缓存和存储系统。由于 Redis 的键包括 string、hash、list、set、sorted set、bitmap 和 hyperloglog，因此常常被称为数据结构服务器。我们可以在这些类型上面执行原子操作，例如，追加字符串，增加哈希中的值，加入一个元素到列表，计算集合的交集、并集和差集，或者是从有序集合中获取最高排名的元素。

为了满足高性能，Redis 采用内存（in-memory）数据集（dataset）。根据使用场景，我们可以通过每隔一段时间转储数据集到磁盘，或者追加每条命令到日志来进行持久化。持久化也可以被禁用，如果我们只是需要一个功能丰富、网络化的内存缓存的话。Redis 还支持主从异步复制、非阻塞初次同步、网络断开时自动重连局部重同步。本章主要介绍 Redis 数据库的基础知识，为后面实战项目中使用 Redis 数据库做好准备。

本章主要知识点：

- Redis 安装
- Redis 常用数据类型
- Redis 排序
- Redis 事务
- Redis 发布订阅及示例
- Redis 持久化

9.1 Redis 环境安装

9.1.1 简介

Redis 是一个开源的 key-value 数据库。它又经常被认为是一个数据结构服务器，因为它的 value 不仅包括基本的 string 类型，还包括 list、set、sorted set 和 hash 类型。当然这些类型的元素也都是

string 类型，也就是说 list、set 这些集合类型也只能包含 string 类型。我们可以在这些类型上执行很多原子性的操作，比如对一个字符 value 追加字符串（APPEND 命令），加加（++）或者减减（--）一个数字字符串（INCR 命令，当然是按整数处理的），还可以对 list 类型进行 push 或者 pop 元素操作（可以模拟栈和队列），对于 set 类型可以进行一些集合相关操作（Intersection Union Difference）。

memcache 也有类似于 ++、-- 的命令。不过 memcache 的 value 只包括 string 类型，远没有 Redis 的 value 类型丰富。和 memcache 一样为了性能，Redis 的数据通常都是存放到内存中的。当然 Redis 可以每间隔一定时间将内存中的数据写入磁盘以防止数据丢失。Redis 也支持主从复制机制（master-slave replication）。Redis 的其他特性包括简单的事务支持和发布订阅（pub/sub）通道功能，而且 Redis 的配置管理非常简单，还提供有各种语言版本的开源客户端类库。

9.1.2　安装

本章采用的是 Redis 6.2.0，其下载地址为：http://redis.io/download。

首先将下载的 Redis 压缩包放到虚拟机 server201（对应 IP 地址为 192.168.56.201）主机上，然后在 Linux 下运行如下命令进行安装：

```
$ tar xzf redis-6.2.0.tar.gz
$ cd redis-6.2.0
$ make
```

make 完后，redis-6.2.0 目录下会出现编译后的 Redis 服务程序 redis-server，还有用于测试的客户端程序 redis-cli。

下面启动 Redis 服务：

```
$ ./redis-server
```

注意这种方式启动 Redis 使用的是默认配置。也可以通过启动参数告诉 Redis 使用指定配置文件启动，命令如下：

```
$ ./redis-server redis.conf
```

redis.conf 是一个默认的配置文件，我们可以根据需要使用自己的配置文件。

启动 Redis 服务进程后，就可以使用测试客户端程序 redis-cli 和 redis 服务交互了，比如：

```
$ ./redis-cli
redis> set foo bar
OK
redis> get foo
"bar"
```

这里演示了使用 get 和 set 命令操作简单类型 value 的例子，foo 是 key，bar 是个 string 类型的 value。

9.1.3　Java 客户端

客户端 JAR 包的下载地址 http://cloud.github.com/downloads/alphazero/jredis 。与 Redis 对应的 jedis JAR 包可以下载 3.7.0 版本。

在 IDEA 或 Eclipse 中新建一个 Java 项目，然后添加 jredis 包引用。下面是个 helloworld 程序。

```java
import org.jredis.*;
import org.jredis.ri.alphazero.JRedisClient;
public class App {
    public static void main(String[] args) {
        try {
            JRedis jr = new JRedisClient("192.168.56.201",6379); //redis服务地址和端口号
            String key = "mKey";
            jr.set(key, "hello,redis!");
            String v = new String(jr.get(key));
            System.out.println(v);
        } catch (Exception e) {
            // TODO: handle exception
        }
    }
}
```

程序运行后，我们可以看到打印结果为"hello,redis!"，表明 Java 程序通过 JredisClient 类访问 Redis 成功。

9.2　Redis 常见数据类型

本节将介绍 Redis 支持的各种常见数据类型，包括 string、list、set、sorted set 和 hash。

9.2.1　key

Redis 本质上一个 key-value 数据库，所以我们首先来看看它的 key。key 也是字符串类型，但是 key 中不能包括边界字符。由于 key 不是 binary safe 的字符串，所以像"my key"和"mykey\n"这样包含空格和换行符的 key 是不被允许的。

在 Redis 内部并不限制使用 binary 字符，这是 Redis 协议限制的。"\r\n"在协议格式中会作为特殊字符。

Redis 1.2 以后的协议中，部分命令已经开始使用新的协议格式了，比如 MSET。总之目前还是把包含边界字符的 key 当作非法的 key。另外关于 key 的一个格式约定需要介绍一下，即 object-type:id:field，比如 user:1000:password、blog:xxidxx:title。

还有 key 的长度最好不要太长，太长会占用较多的内存，而且查找速度也更慢。所以推荐使用相对较短的 key。但过短的 key 也同样不推荐，因为可读性差，比如 u:1000:pwd 显然没上面的 user:1000:password 可读性好。

下面介绍 key 的相关命令：

（1）exits key：测试指定 key 是否存在，返回 1 表示存在，0 表示不存在。

（2）del key1 key2 ... keyN：删除给定 key，返回删除 key 的数目，0 表示给定 key 都不存在。

（3）type key：返回给定 key 的 value 类型。返回 none 表示不存在 key，string 表示字符类型、list 表示链表类型、set 表示无序集合类型等。

（4）keys pattern：返回匹配指定模式的所有 key。示例如下：

```
redis> set test dsf
OK
redis> set tast dsaf
OK
redis> set tist adff
OK
redis> keys t*
1. "tist"
2. "tast"
3. "test"
redis> keys t[ia]st
1. "tist"
2. "tast"
redis> keys t?st
1. "tist"
2. "tast"
3. "test"
```

（5）randomkey：返回从当前数据库中随机选择的一个 key，如果当前数据库是空的，则返回空串。

（6）rename oldkey newkey：原子地重命名一个 key，如果 newkey 存在，则它将会被覆盖。返回 1 表示成功，0 表示失败。失败可能是因为 oldkey 不存在或者和 newkey 相同。

（7）renamenx oldkey newkey：作用同上，区别在于如果 newkey 存在，则返回失败。

（8）dbsize：返回当前数据库的 key 数量。

（9）expire key seconds：为 key 指定过期时间，单位是秒。返回 1 表示成功，0 表示 key 已经设置了过期时间或者不存在。

（10）ttl key：返回设置了过期时间的 key 的剩余过期秒数，-1 表示 key 不存在或者没有设置过期时间。

（11）select db-index：通过索引选择数据库，默认连接的数据库索引是 0，默认的数据库数量是 16 个。返回 1 表示成功，0 表示失败。

（12）move key db-index：将 key 从当前数据库移动到指定数据库。返回 1 表示成功，0 表示 key 不存在，或者已经在指定数据库中。

（13）flushdb：删除当前数据库中的所有 key。此方法不会失败。

（14）flushall：删除所有数据库中的所有 key。此方法不会失败。

9.2.2　string 类型

string 是 Redis 最基本的类型，而且 string 类型是二进制安全的。意思是 Redis 的 string 可以包含任何数据，比如 JPG 图片或者序列化的对象。

从内部实现来看，string 其实可以看作 byte 数组，最大上限是 1GB。下面是 string 类型的定义：

```
struct sdshdr {
    long len;
    long free;
    char buf[];
};
```

其中，buf 是个 char 数组，用于存储实际的字符串内容，其实 char 和 C#中的 byte 是等价的，都是一个字节；len 是 buf 数组的长度，free 是数组中剩余可用字节数。由此可以理解为什么 string 类型是二进制安全的了，因为它本质上就是个 byte 数组，因此 string 可以包含任何数据。另外 string 类型可以被部分命令按 int 处理，比如 incr 等命令，下面将详细介绍。还有 Redis 的其他类型，例如 list、set、sorted set、hash，它们包含的元素都只能是 string 类型。

如果只使用 string 类型，Redis 就可以被看作加上持久化特性的 memcached。当然 Redis 对 string 类型的操作比 memcached 多很多，方法如下：

（1）set key value：设置 key 对应的值为 string 类型的 value，返回 1 表示成功，0 表示失败。

（2）setnx key value：作用同上，如果 key 已经存在，则返回 0 。nx 是 not exist 的意思。

（3）get key：获取 key 对应的 string 值，如果 key 不存在，则返回 nil。

（4）getset key value：原子地设置 key 的值，并返回 key 的旧值。如果 key 不存在，则返回 nil。

（5）mget key1 key2 ... keyN：一次获取多个 key 的值，如果对应 key 不存在，则对应返回 nil。

下面示例首先清空当前数据库，然后设置 k1、k2，获取 k3 时对应返回 nil。

```
redis> flushdb
OK
redis> dbsize
(integer) 0
redis> set k1 a
OK
redis> set k2 b
OK
redis> mget k1 k2 k3
1. "a"
2. "b"
3. (nil)
```

（6）mset key1 value1 ... keyN valueN：一次设置多个 key 的值。成功则返回 1，表示所有的值都设置了；失败则返回 0，表示没有任何值被设置。

（7）msetnx key1 value1 ... keyN valueN：作用同上，但是不会覆盖已经存在的 key。

（8）incr key：对 key 的值做加加操作，并返回新的值。注意，incr 一个不是 int 的 value 将会返回错误；incr 一个不存在的 key，则设置 key 为 1。

（9）decr key：作用同上，但是做的是减减操作。decr 一个不存在 key，则设置 key 为-1

（10）incrby key integer：作用同 incr，加指定值。key 不存在时候会设置 key，并认为原来的 value 是 0。

（11）decrby key integer：作用同 decr，减指定值。decrby 完全是为了可读性，我们完全可以通过 incrby 一个负值来实现同样的效果，反之一样。

（12）append key value：给指定 key 的字符串值追加 value，返回新字符串值的长度。

示例如下：

```
redis> set k hello
OK
redis> append k ,world
(integer) 11
redis> get k
"hello,world"
```

（13）substr key start end：返回截取过的 key 的字符串值，注意并不修改 key 的值。下标是从 0 开始的，断续使用上面的例子：

```
redis> substr k 0 8
"hello,wor"
redis> get k
"hello,world"
```

9.2.3　list

Redis 列表仅仅是按照插入顺序排序的字符串列表。可以添加一个元素到 Redis 列表的头部（左边）或者尾部（右边）。

LPUSH 命令用于插入一个元素到列表的头部，RPUSH 命令用于插入一个元素到列表的尾部。当这两个命令操作一个不存在的键时，将会创建一个新的列表。同样，如果一个操作会清空列表，那么该键将会从键空间（key space）中移除。这些都是非常方便的语义，因为列表命令如果使用不存在的键作为参数，就会表现得像命令运行在一个空列表上一样。

Redis 的 list 类型其实就是一个每个子元素都是 string 类型的双向链表，因此[lr]push 和[lr]pop 命令的算法时间复杂度都是 O(1)。

另外，list 会记录链表的长度，因此 llen 操作也是 O(1)。链表的最大长度是 $2^{32}-1$。我们可以通过 push、pop 操作从链表的头部或者尾部添加或删除元素，这使得 list 既可以用作栈，也可以用作队列。list 的 pop 操作还有阻塞版本。当我们[lr]pop 一个 list 对象时，如果 list 为空，或者不存在，则会立即返回 nil。但是阻塞版本的 b[lr]pop 则可以被阻塞，当然也可以加超时时间，超时后也会返回 nil。

为什么要阻塞版本的 pop 呢？主要是为了避免轮询。举个简单的例子，如果我们用 list 来实现一个工作队列，执行任务的 thread 可以调用阻塞版本的 pop 去获取任务，这样就可以避免轮询去检查是否有任务存在。当任务来时，工作线程可以立即返回，也可以避免轮询带来的延迟。

下面介绍 list 的相关命令。

（1）lpush key string：在 key 对应的 list 的头部添加字符串元素，返回 1 表示成功，0 表示 key 存在且不是 list 类型。

（2）rpush key string：作用同上，在尾部添加。

（3）llen key：返回 key 对应的 list 的长度，key 不存在则返回 0，如果 key 对应类型不是 list 则返回错误。

（4）lrange key start end：返回指定区间内的元素，下标从 0 开始，负值表示从后面开始计算，-1 表示倒数第一个元素。key 不存在则返回空列表。

（5）ltrim key start end：截取 list，保留指定区间内的元素，成功则返回 1，key 不存在则返回错误。

（6）lset key index value：设置 list 中指定下标的元素值，成功则返回 1，key 或者下标不存在则返回错误。

（7）lrem key count value：从 key 对应的 list 中删除 count 个和 value 相同的元素。count 为 0 时删除全部。

（8）lpop key：从 list 的头部删除元素，并返回删除元素。如果 key 对应的 list 不存在或者为空，则返回 nil；如果 key 对应值不是 list，则返回错误。

（9）rpop：作用同上，但是从尾部删除。

（10）blpop key1...keyN timeout：从左到右对第一个非空 list 进行 lpop 操作并返回删除的元素，比如 blpop list1 list2 list3 0，如果 list1 不存在，list2 和 list3 都是非空，则对 list2 做 lpop 并返回从 list2 中删除的元素；如果所有的 list 都为空或不存在，则会阻塞 timeout 秒，timeout 为 0 表示一直阻塞。

当阻塞时，如果有 client 对 key1...keyN 中的任意 key 进行 push 操作，则第一个在这个 key 上被阻塞的 client 会立即返回。如果超时发生，则返回 nil。这有点像 Unix 的 select 或者 poll。

（11）brpop：作用同 blpop，只不过一个是从头部删除，一个是从尾部删除。

（12）rpoplpush srckey destkey：从 srckey 对应的 list 的尾部移除元素并添加到 destkey 对应的 list 的头部，最后返回被移除的元素值，整个操作是原子的。如果 srckey 为空或者不存在，则返回 nil。

下面是关于 list 的使用的示例。

```
redis 127.0.0.1:6379> DEL runoob
redis 127.0.0.1:6379> lpush runoob redis
(integer) 1
redis 127.0.0.1:6379> lpush runoob mongodb
(integer) 2
redis 127.0.0.1:6379> lpush runoob rabbitmq
(integer) 3
redis 127.0.0.1:6379> lrange runoob 0 10
1. "rabbitmq"
2. "mongodb"
3. "redis"
```

9.2.4　set

Redis 集合是没有顺序的字符串集合（collection）。可以在 O(1)的时间复杂度内添加或删除元素，以及测试元素存在与否（不管集合中有多少元素都是常量时间）。

Redis 集合具有不允许成员重复的性质。多次加入同一个元素到集合，最终也只会有一个在其中。实际上，这意味着加入一个元素到集合中并不需要检查该元素是否已经存在。

Redis 的 set 是 string 类型的无序集合。set 元素最大可以包含 $2^{32}-1$ 个元素。set 是通过 hash table 实现的，所以添加、删除、查找的复杂度都是 O(1)。hash table 会随着添加或者删除而自动地调整大小。需要注意的是，调整 hash table 大小时，需要同步（获取写锁）操作，因此会阻塞其他读写操作。

关于 set 集合类型，除了基本的添加和删除操作外，其他有用的操作还包含集合的取并集（union）、交集（intersection）、差集（difference）。通过这些操作可以很容易实现 SNS 中的好友推荐和 Blog 的 tag 功能。

下面详细介绍 set 的相关命令。

（1）sadd key member：添加一个 string 元素到 key 对应的 set 集合中，成功则返回 1；如果元素已经在集合中，则返回 0。key 对应的 set 不存在返回错误。

（2）srem key member：从 key 对应的 set 中移除给定元素，成功则返回 1；如果 member 在集合中不存在或者 key 不存在，则返回 0；如果 key 对应的不是 set 类型的值，则返回错误。

（3）spop key：删除并返回 key 对应的 set 中随机的一个元素，如果 set 为空或者 key 不存在，则返回 nil。

（4）srandmember key：作用同 spop，随机取 set 中的一个元素，但是不删除元素。

（5）smove srckey dstkey member：从 srckey 对应的 set 中移除 member 并添加到 dstkey 对应的 set 中，整个操作是原子的，成功则返回 1；如果 member 在 srckey 中不存在，则返回 0；如果 key 不是 set 类型，则返回错误。

（6）scard key：返回 set 的元素个数，如果 set 为空或者 key 不存在，则返回 0。

（7）sismember key member：判断 member 是否在 set 中，存在则返回 1，0 表示不存在或者 key 不存在。

（8）sinter key1 key2...keyN：返回所有给定 key 的交集。

（9）sinterstore dstkey key1...keyN：作用同 sinter，但是会同时将交集存到 dstkey 下。

（10）sunion key1 key2...keyN：返回所有给定 key 的并集。

（11）sunionstore dstkey key1...keyN：作用同 sunion，但是会同时保存并集到 dstkey 下。

（12）sdiff key1 key2...keyN：返回所有给定 key 的差集。

（13）sdiffstore dstkey key1...keyN：作用同 sdiff，但是会同时保存差集到 dstkey 下。

（14）smembers key：返回 key 对应的 set 的所有元素，结果是无序的。

下面是关于 set 的使用示例：

```
redis 127.0.0.1:6379> DEL runoob
redis 127.0.0.1:6379> sadd runoob redis
(integer) 1
redis 127.0.0.1:6379> sadd runoob mongodb
(integer) 1
redis 127.0.0.1:6379> sadd runoob rabbitmq
(integer) 1
redis 127.0.0.1:6379> sadd runoob rabbitmq
```

```
(integer) 0
redis 127.0.0.1:6379> smembers runoob

1. "redis"
2. "rabbitmq"
3. "mongodb"
```

注意：以上示例中 rabbitmq 添加了两次，但根据集合内元素的唯一性，第二次插入的元素将被忽略。

9.2.5 sorted set

sorted set 和 set 极为相似，它们都是字符串的集合，都不允许重复的成员出现在同一个 set 中。它们之间的主要差别是，sorted set 中的每一个成员都会有一个分数（score）与之关联，Redis 正是通过分数来为集合中的成员进行从小到大的排序。然而，需要额外指出的是，尽管 sorted set 中的成员是唯一的，但是分数（score）却是可以重复的。在 sorted set 中添加、删除或更新一个成员都是非常快速的操作，其时间复杂度为集合中成员数量的对数（二分查找）。由于 sorted set 中的成员在集合中的位置是有序的，因此，即便是访问位于集合中部的成员，也仍然非常高效。事实上，Redis 所具有的这一特征，在很多其他类型的数据库中是很难实现的，换句话说，在该特征上要想达到和 Redis 同样的高效，在其他数据库中进行建模是非常困难的。

sorted set 是 skip list 和 hash table 的混合体。当元素被添加到集合中时，一个元素到 score 的映射被添加到 hash table 中，因此给定一个元素获取 score 的时间复杂度是 O(1)。sorted set 最常用的使用方式是作为索引来使用。我们可以把要排序的字段作为 score 存储，对象的 id 作为元素存储。

sorted set 的相关命令如下：

（1）zadd key score member：添加元素到集合，若元素在集合中存在，则更新对应的 score。

（2）zrem key member：删除指定元素，返回 1 表示成功，如果元素不存在则返回 0。

（3）zincrby key incr member：增加对应 member 的 score 值，然后移动元素并保持 skip list 有序。返回更新后的 score 值。

（4）zrank key member：返回指定元素在集合中的排名（下标），集合中元素是按 score 从小到大排序的。

（5）zrevrank key member：作用同上，但是集合中的元素是按 score 从大到小排序的。

（6）zrange key start end：类似 lrange 操作，从集合中去指定区间的元素。返回的是有序结果。

（7）zrevrange key start end：作用同上，返回结果是按 score 逆序排列的。

（8）zrangebyscore key min max：返回集合中 score 在给定区间的元素。

（9）zcount key min max：返回集合中 score 在给定区间的数量。

（10）zcard key：返回集合中元素的个数。

（11）zscore key element：返回给定元素对应的 score。

（12）zremrangebyrank key min max：删除集合中排名在给定区间的元素。

（13）zremrangebyscore key min max：删除集合中 score 在给定区间的元素。

下面是关于 sorted set 的使用示例：

```
redis 127.0.0.1:6379> ZADD runoobkey 1 redis
(integer) 1
redis 127.0.0.1:6379> ZADD runoobkey 2 mongodb
(integer) 1
redis 127.0.0.1:6379> ZADD runoobkey 3 mysql
(integer) 1
redis 127.0.0.1:6379> ZADD runoobkey 3 mysql
(integer) 0
redis 127.0.0.1:6379> ZADD runoobkey 4 mysql
(integer) 0
redis 127.0.0.1:6379> ZRANGE runoobkey 0 10 WITHSCORES
1. "redis"
2. "1"
3. "mongodb"
4. "2"
5. "mysql"
6. "4"
```

在以上示例中我们通过命令 ZADD 向 Redis 的有序集合中添加了 3 个值并关联上分数。

9.2.6 hash

Redis hash 是一个 string 类型的 field 和 value 的映射表，它的添加、删除操作的时间复杂度都是 O(1)（平均）。hash 特别适用于存储对象。相较于将对象的每个字段存储为单个 string 类型，将一个对象存储在 hash 类型中会占用更少的内存，并且可以更方便地存取整个对象。省内存的原因是新建一个 hash 对象时时开始是用 zipmap（又称为 small hash）来存储的。这个 zipmap 其实并不是 hash table，但是 zipmap 相较于正常的 hash 实现，可以节省不少 hash 本身需要的一些元数据存储开销。尽管 zipmap 的添加、删除、查找的时间复杂度都是 O(n)，但是由于一般对象的 field 数量都不太多，因此使用 zipmap 也是很快的，也就是说添加、删除操作的平均的时间复杂度还是 O(1)。当 field 或者 value 的大小超出一定限制后，Redis 会在内部自动将 zipmap 替换成正常的 hash 实现。这个限制可以在配置文件中指定。比如：

```
hash-max-zipmap-entries 64 #配置字段最多 64 个
hash-max-zipmap-value 512 #配置 value 最大为 512 字节
```

hash 的相关命令如下：

（1）hset key field value：设置 hash field 为指定值，如果 key 不存在，则先创建。

（2）hget key field：获取指定的 hash field。

（3）hmget key field1 ... fieldN：获取全部指定的 hash field。

（4）hmset key field1 value1 ... fieldN valueN：同时设置 hash 的多个 field。

（5）hincrby key field integer：将指定的 hash field 加上给定值。

（6）hexists key field：测试指定 field 是否存在。

（7）hdel key field：删除指定的 hash field。

（8）hlen key：返回指定 hash 的 field 数量。

(9) hkeys key：返回 hash 的所有 field。

(10) hvals key：返回 hash 的所有 value。

(11) hgetall：返回 hash 的所有 filed 和 value。

下面是关于 hash 的使用示例：

```
redis 127.0.0.1:6379> DEL runoob
redis 127.0.0.1:6379> HMSET runoob field1 "Hello" field2 "World"
"OK"
redis 127.0.0.1:6379> HGET runoob field1
"Hello"
redis 127.0.0.1:6379> HGET runoob field2
"World"
```

DEL runoob 用于删除前面测试使用的 key。上面示例中我们使用了 Redis HMSET、HGET 命令，HMSET 设置了两个 field-value 对，HGET 获取 field 对应的 value。

9.3　Redis 排序

在了解完各种 Redis 类型后，本节将介绍 Redis 排序命令。Redis 支持对 list、set 和 sorted set 元素进行排序，排序命令是 sort ，完整的命令格式如下：

```
sort key [BY pattern] [LIMIT start count] [GET pattern] [ASC|DESC] [ALPHA] [STORE dstkey]
```

下面我们一一说明各个命令选项。

1. sort key

这个是最简单的情况，没有任何选项就是简单地对集合自身元素进行排序并返回排序结果。示例如下：

```
redis> lpush ml 12
(integer) 1
redis> lpush ml 11
(integer) 2
redis> lpush ml 23
(integer) 3
redis> lpush ml 13
(integer) 4
redis> sort ml
1. "11"
2. "12"
3. "13"
4. "23"
```

2. [BY pattern]

除了可以按照集合元素自身值排序外，还可以将集合元素内容按照给定 pattern 组合成新的 key，并按照新 key 中对应的内容进行排序。下面的例子接着使用上述例子中的 ml 集合做演示：

```
redis> set name11 nihao
OK
redis> set name12 wo
OK
redis> set name13 shi
OK
redis> set name23 lala
OK
redis> sort ml by name*
1. "13"
2. "23"
3. "11"
4. "12"
```

*代表了 ml 中的元素值，所以本示例排序是按照 name11、name12、name13、name23 这 4 个 key 对应值排序的，当然返回的还是排序后 ml 集合中的元素。

3. [GET pattern]

上面的例子返回的都是 ml 集合中的元素。我们也可以通过 get 选项去获取指定 pattern 作为新 key 对应的值。查看如下组合起来的例子：

```
redis> sort ml by name* get name* alpha
1. "lala"
2. "nihao"
3. "shi"
4. "wo"
```

这次返回的就不再是 ml 中的元素了，而是 name11、name12、name13、name23 对应的值。当然排序是按照 name11、name12、name13、name23 的值并根据字母顺序排列的。另外 get 选项可以有多个。查看如下例子（#特殊符号引用的是原始集合，也就是 ml）：

```
redis> sort ml by name* get name* get # alpha
1. "lala"
2. "23"
3. "nihao"
4. "11"
5. "shi"
6. "13"
7. "wo"
8. "12"
```

最后还有一个引用 hash 类型字段的特殊字符 "->"，其使用示例如下：

```
redis> hset user1 name hanjie
(integer) 1
redis> hset user11 name hanjie
(integer) 1
redis> hset user12 name 86
(integer) 1
```

```
redis> hset user13 name lxl
(integer) 1
redis> sort ml get user*->name
1. "hanjie"
2. "86"
3. "lxl"
4. (nil)
```

很容易理解。注意，当对应的 user23 不存在时，返回的是 nil。

4. [ASC|DESC] [ALPHA]

sort 默认的排序方式是从小到大排序（asc），也可以按照逆序或者按字符顺序排序。逆序可以加上 desc 选项，想按字母顺序排序可以加上 alpha 选项，当然 alpha 可以和 desc 一起用。下面是个按字母顺序排序的例子：

```
redis> lpush mylist baidu
(integer) 1
redis> lpush mylist hello
(integer) 2
redis> lpush mylist xhan
(integer) 3
redis> lpush mylist soso
(integer) 4
redis> sort mylist
1. "soso"
2. "xhan"
3. "hello"
4. "baidu"
redis> sort mylist alpha
1. "baidu"
2. "hello"
3. "soso"
4. "xhan"
redis> sort mylist desc alpha
1. "xhan"
2. "soso"
3. "hello"
4. "baidu"
```

5. [LIMIT start count]

上面例子返回的都是全部结果，limit 选项可以限定返回结果的数量。示例如下：

```
redis> sort ml get name* limit 1 2
1. "wo"
2. "shi"
```

start 下标是从 0 开始的，这里的 limit 选项的意思是从第 2 个元素开始获取 2 个。

6. [STORE dstkey]

如果集合经常按照固定的模式去排序，那么把排序结果缓存起来会减少不少 CPU 开销。使用 store 选项可以将排序内容保存到指定 key 中，保存的类型是 list，示例如下：

```
redis> sort ml get name* limit 1 2 store cl
(integer) 2
redis> type cl
list
redis> lrange cl 0 -1
1. "wo"
2. "shi"
```

这个例子中，我们将排序结果保存到了 cl 中。

功能介绍完后，再讨论一下关于排序的一些问题。如果我们有多个 Redis server，不同的 key 可能存在于不同的 server 上，比如 name11、name12、name13、name23 很有可能分别在 4 个不同的 server 上存储着，这种情况会对排序性能造成很大的影响。Redis 作者在他的 Blog 上提到了这个问题的解决办法，就是通过 key tag 将需要排序的 key 都放到同一个 server 上 。由于具体决定哪个 key 存在哪个服务器上，一般都是在 client 端使用 hash 办法来实现的，因此我们可以只对 key 的部分进行 hash。举个例子，假如我们的 client 端发现 key 中包含[]，那么只对 key 中[]包含的内容进行 hash。我们将 4 个 name 相关的 key 都做这样的命名——[name]12、[name]13、[name]23、[name]23，于是 client 程序就会把它们都放到同一 server 上。

还有一个问题也比较严重：如果要 sort 的集合非常大的话，排序就会消耗很长时间。由于 Redis 是单线程的，所以长时间的排序操作会阻塞其他 client 的请求。解决办法是通过主从复制机制将数据复制到多个 slave 上。然后我们只在 slave 上做排序操作，并尽可能地对排序结果做缓存。另外还有一个方案就是采用 sorted set 对需要按某个顺序访问的集合建立索引。

9.4 Redis 事务

Redis 对事务的支持目前还比较简单。Redis 只能保证一个 client 发起的事务中的命令可以连续地执行，而中间不会插入其他 client 的命令。 由于 Redis 是单线程来处理所有 client 请求的，因此要做到这点很容易。一般情况下 Redis 在接收到一个 client 发来的命令后，会立即处理并返回处理结果，但是当一个 client 在一个连接中发出 multi 命令后，这个连接会进入一个事务上下文，该连接后续的命令并不会立即执行，而是先放到一个队列中。当从此连接收到 exec 命令后，Redis 会顺序地执行队列中的所有命令，并将所有命令的运行结果打包到一起返回给 client，然后此连接就结束事务上下文。下面可以看一个例子：

```
redis> multi
OK
redis> incr a
QUEUED
redis> incr b
QUEUED
```

```
redis> exec
1. (integer) 1
2. (integer) 1
```

从这个例子中我们可以看到，incr a、incr b 命令发出后并没有执行，而是被放到了队列中。调用 exec 后两条命令被连续地执行，最后返回的是两条命令执行后的结果。

我们可以调用 discard 命令来取消一个事务，接着上面的例子：

```
redis> multi
OK
redis> incr a
QUEUED
redis> incr b
QUEUED
redis> discard
OK
redis> get a
"1"
redis> get b
"1"
```

可以发现这次 incr a、incr b 都没被执行。discard 命令其实就是清空事务的命令队列，并退出事务上下文。

虽说 Redis 事务在本质上也相当于序列化隔离级别的了，但是由于事务上下文的命令只排队并不立即执行，因此事务中的写操作不能依赖事务中的读操作结果。看下面例子：

```
redis> multi
OK
redis> get a
QUEUED
redis> get b
QUEUED
redis> exec
1. "1"
2. "1"
```

假如我们想用事务来实现 incr 操作，可以如下这样做吗？

```
redis> get a
"1"
redis> multi
OK
redis> set a 2
QUEUED
redis> exec
1. OK
redis> get a,
"2"
```

结论很明显，这样是不行的，这样操作和先 get a 然后直接 set a 是没有区别的。get a 和 set a 并不能保证两个命令是连续执行的（get 操作不在事务上下文中），很可能有两个 client 同时在做这个操作。我们期望的结果是加两次 a 从原来的 1 变成 3，但是很有可能两个 client 的 get a 取到的都是 1，造成最终加两次结果却是 2。主要问题是我们没有对共享资源 a 的访问进行任何的同步操作，也就是说 Redis 没提供任何的加锁机制来同步对 a 的访问。

Redis 2.1 后添加了 watch 命令，可以用来实现乐观锁。下面是正确实现 incr 命令的例子，在操作开始加了 watch a：

```
redis> watch a
OK
redis> get a
"1"
redis> multi
OK
redis> set a 2
QUEUED
redis> exec
1. OK
redis> get a,
"2"
```

watch 命令会监视给定的 key，当执行 exec 的时候，如果监视的 key 从调用 watch 后发生过变化，则整个事务会失败。也可以调用 watch 多次监视多个 key，这样就可以对指定的 key 加乐观锁了。注意，watch 的 key 是对整个连接有效的，事务也一样。如果连接断开，则监视和事务都会被自动清除。当然，exec、discard、unwatch 命令都会清除连接中的所有监视。

Redis 的事务实现比较简单，当然会存在一些问题。第一个问题是 Redis 只能保证事务的每个命令连续执行，如果事务中的一个命令失败了，它并不回滚其他命令。比如使用的命令类型不匹配：

```
redis> set a 5
OK
redis> lpush b 5
(integer) 1
redis> set c 5
OK
redis> multi
OK
redis> incr a
QUEUED
redis> incr b
QUEUED
redis> incr c
QUEUED
redis> exec
1. (integer) 6
2. (error) ERR Operation against a key holding the wrong kind of value
3. (integer) 6
```

可以看到虽然 incr b 执行失败了，但是其他两个命令还是执行成功了。

9.5 Redis 发布订阅及示例

1. 发布订阅

Redis 发布订阅（pub/sub）是一种消息通信模式：发送者（pub）发送消息，订阅者（sub）接收消息。Redis 客户端可以订阅任意数量的频道。

图 9-1 展示了频道 channel1 以及订阅这个频道的 3 个客户端——client2、client5 和 client1 之间的关系。

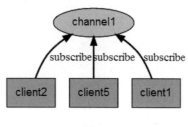

图 9-1

当有新消息通过 PUBLISH 命令发送给频道 channel1 时，这个消息就会被发送给订阅它的 3 个客户端，如图 9-2 所示。

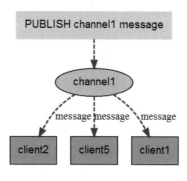

图 9-2

发布订阅是一种消息通信模式，主要目的是解耦消息发布者和消息订阅者之间的耦合，这一点和设计模式中的观察者模式比较相似。pub/sub 不仅解决发布者和订阅者直接代码级别的耦合，也解决两者在物理部署上的耦合。Redis 作为一个 pub/sub server，在订阅者和发布者之间起到了消息路由的功能。订阅者可以通过 subscribe 和 psubscribe 命令向 Redis server 订阅自己感兴趣的消息类型，Redis 将消息类型称为通道（channel）。当发布者通过 publish 命令向 Redis server 发送特定类型的消息时，订阅该消息类型的全部 client 都会收到此消息。这里消息的传递是多对多的，一个 client 可以订阅多个 channel，也可以向多个 channel 发送消息。

2. 发布订阅示例

这里使用两个不同的 client，一个是 Redis 自带的 redis-cli，另一个是用 Java 编写的简单的 client。代码如下：

```java
import java.net.*;
import java.io.*;
public class PubSubTest {
    public static void main(String[] args) {
        String cmd = args[0]+"\r\n";
        try {
            Socket socket = new Socket("192.168.56.55",6379);
            InputStream in = socket.getInputStream();
            OutputStream out = socket.getOutputStream();
            out.write(cmd.getBytes());  //发送订阅命令
            byte[] buffer = new byte[1024];
            while (true) {
                int readCount = in.read(buffer);
                System.out.write(buffer, 0, readCount);
                System.out.println("--------------------------------------");
            }
        } catch (Exception e) {
        }
    }
}
```

代码就是简单地从命令行读取传送过来的订阅命令,然后通过一个 socket 连接发送给 Redis server,再进入 while 循环一直读取 Redis server 传送过来的订阅消息,并打印到控制台。

步骤01 首先编译并运行此 Java 程序:

```
javac PubSubTest.java
java PubSubTest "subscribe news.share news.blog"
*3
$9
subscribe
$10
news.share
:1

*3
$9
subscribe
$9
news.blog
:2
--------------------------------------
```

步骤02 启动 redis-cli:

```
redis> psubscribe news.*
Reading messages... (press Ctrl-c to quit)
1. "psubscribe"
2. "news.*"
3. (integer) 1
```

步骤03 再启动一个 redis-cli 用来发布两条消息：

```
redis> publish news.share "share a link http://www.google.com"
(integer) 2
redis> publish news.blog "I post a blog"
(integer) 2
```

步骤04 查看两个订阅 client 的输出。

此时 Java client 打印如下内容：

```
*3
$7
message
$10
news.share
$34
share a link http://www.google.com
-------------------------------------
*3
$7
message
$9
news.blog
$13
I post a blog
-------------------------------------
```

另一个 redis-cli 输出如下：

```
1) "pmessage"
2) "news.*"
3) "news.share"
4) "share a link http://www.google.com"

1) "pmessage"
2) "news.*"
3) "news.blog"
4) "I post a blog"
```

分析如下：

Java client 使用 subscribe 命令订阅 news.share 和 news.blog 两个通道，然后立即收到 server 返回的订阅成功消息，可以看出 Redis 的协议是文本类型的，这里不解释具体协议内容，读者可以参考 http://redis.io/topics/protocol 或者 http://terrylee.me/blog/post/2011/01/26/redis-internal-part3.aspx。这个报文内容有两部分，第一部分表示该 socket 连接使用 subscribe 订阅 news.share 成功后，此连接订阅通道数为 1；第二部分表示使用 subscribe 订阅 news.blog 成功后，该连接订阅通道总数为 2。

Redis client 使用 psubscribe 订阅了一个使用通配符的通道（*表示任意字符串），此订阅会收到所有与 news.* 匹配的通道消息。redis-cli 打印到控制台的订阅成功消息，表示使用 psubscribe 命令订阅 news.* 成功后，连接订阅通道总数为 1。

当我们在一个 client 使用 publish 向 news.share 和 news.blog 通道发出两个消息后。Redis 返回的 (integer) 2 表示有两个连接收到了此消息。这个可以通过观察两个订阅者的输出进行验证。

看完这个示例后，读者应该对 pub/sub 功能有了一定认识。需要注意的是，当一个连接通过 subscribe 或者 psubscribe 订阅通道后就进入订阅模式。在这种模式下，除了再订阅额外的通道或者用 unsubscribe 或者 punsubscribe 命令退出订阅模式外，不能再发送其他命令。另外，使用 psubscribe 命令订阅多个通配符通道，如果一个消息匹配上了多个通道模式，则会多次收到同一个消息。

9.6 Redis 持久化

Redis 是一个支持持久化的内存数据库，也就是说 Redis 需要经常将内存中的数据同步到磁盘来保证持久化。Redis 支持两种持久化方式：一种是 Snapshotting（快照），也是默认方式；另一种是 Append-only file（缩写 aof）的方式。本节分别介绍这两种持久化方式。

1. Snapshotting

快照是默认的持久化方式。这种方式是就是将内存中的数据以快照的方式写入二进制文件中，默认的文件名为 dump.rdb。我们可以通过配置设置自动做快照持久化。可以配置 Redis 在 n 秒内如果超过 m 个 key 被修改就自动做快照，下面是默认的快照保存配置示例：

```
save 900 1    #900 秒内如果超过 1 个 key 被修改，就发起快照保存
save 300 10   #300 秒内如果超过 10 个 key 被修改，就发起快照保存
save 60 10000
```

下面介绍详细的快照保存过程：

（1）Redis 调用 fork 后，有了子进程和父进程。

（2）父进程继续处理 client 请求，子进程负责将内存内容写入临时文件。由于 os 的写时复制机制（copy on write），父子进程会共享相同的物理页面。当父进程处理写请求时，os 会为父进程要修改的页面创建副本，而不是写共享的页面。因此，子进程的地址空间内的数据是 fork 时刻整个数据库的一个快照。

（3）当子进程将快照写入临时文件后，用临时文件替换原来的快照文件，然后子进程退出。

client 也可以使用 save 或者 bgsave 命令通知 Redis 做一次快照持久化。save 操作是在主线程中保存快照的，由于 Redis 用一个主线程来处理所有 client 的请求，因此这种方式会阻塞所有 client 请求，所以不推荐使用。另一点需要注意的是，每次快照持久化都是将内存数据完整写入磁盘一次，并不是增量的，只同步脏数据。如果数据量大而且写操作比较多，则必然会引起大量的磁盘 I/O 操作，可能会严重影响性能。

另外，由于快照方式是在一定间隔时间做一次的，因此如果 Redis 意外宕掉，就会丢失最后一次快照后的所有修改。如果应用要求不能丢失任何修改，则可以采用 aof 持久化方式。

2. Append-only file

aof 比快照方式有更好的持久化性，是因为在使用 aof 持久化方式时，Redis 会将每一个收到的

写命令都通过 write 函数追加到文件中（默认是 appendonly.aof）。当 Redis 重启时，会通过重新执行文件中保存的写命令来在内存中重建整个数据库的内容。当然，由于 os 会在内核中缓存 write 做的修改，因此可能不是立即写到磁盘上，这样 aof 方式的持久化也还是有可能会丢失部分修改。不过可以通过配置文件告诉 Redis，我们想要通过 fsync 函数强制 os 写入磁盘的时机，有如下 3 种方式（默认每秒 fsync 一次）：

```
    appendonly yes              //启用 aof 持久化方式
    # appendfsync always        //每次收到写命令就立即强制写入磁盘，最慢，但是保证完全的持久化，
不推荐使用
    appendfsync everysec        //每秒强制写入磁盘一次，在性能和持久化方面做了很好的折中，推荐
    # appendfsync no            //完全依赖 os，性能最好，持久化没保证
```

aof 的方式也带来了另一个问题：持久化文件会变得越来越大。例如，我们调用 incr test 命令 100 次，文件中必须保存全部的 100 条命令，其实有 99 条都是多余的，因为要恢复数据库的状态，其实文件中保存一条 set test 100 就够了。为了压缩 aof 的持久化文件，Redis 提供了 bgrewriteaof 命令，收到此命令后，Redis 使用与快照类似的方式将内存中的数据以命令的方式保存到临时文件中，最后替换原来的文件。具体过程如下：

（1）Redis 调用 fork，现在有父、子两个进程。

（2）子进程根据内存中的数据库快照，往临时文件中写入重建数据库状态的命令。

（3）父进程继续处理 client 请求，除了把写命令写入原来的 aof 文件中，同时把收到的写命令缓存起来，这样就能保证即使子进程重写失败也不会出问题。

（4）当子进程把快照内容以命令方式写入临时文件中后，子进程发信号通知父进程，然后父进程把缓存的写命令也写入临时文件。

（5）现在父进程可以使用临时文件替换旧的 aof 文件并重命名，后面收到的写命令也开始往新的 aof 文件中追加。

需要注意的是，重写 aof 文件的操作并没有读取旧的 aof 文件，而是将整个内存中的数据库内容用命令的方式重新写入一个新的 aof 文件中，这一点和快照有点类似。

第 10 章

广告点击实时大数据分析项目实战

本实战项目使用 DStream 来实时分析处理用户对广告的点击行为数据。本项目实战首先模拟生成用户广告点击信息，包括时间戳、区域、城市、用户 ID、广告 ID、日期、时间等。然后通过 Kafka 读取广告点击信息，并通过创建 DStream 返回接收到的输入数据。然后基于 DStream 数据统计每天每地区热门广告前 3 位，统计各广告最近 1 小时内的点击量趋势——各广告最近 1 小时内各分钟的点击量等。最后，将统计结果存储到 Redis 数据库中。

本章主要知识点：

- 利用 Kafka 存储和读取实时广告点击数据
- DStream 接收数据
- DStream 数据统计

10.1 项目环境准备

本项目的 Spark 环境依然采用本地模式，只需要引入 Spark 相关 JAR 包依赖即可，不需要额外安装 Spark 环境。

Spark 相关依赖代码如下：

```xml
<dependencies>
    <dependency>
        <groupId>org.apache.spark</groupId>
        <artifactId>spark-core_2.12</artifactId>
        <version>3.0.0</version>
    </dependency>
    <dependency>
        <groupId>org.apache.spark</groupId>
        <artifactId>spark-sql_2.12</artifactId>
```

```xml
        <version>3.0.0</version>
    </dependency>
    <dependency>
        <groupId>org.apache.spark</groupId>
        <artifactId>spark-streaming_2.12</artifactId>
        <version>3.0.0</version>
    </dependency>
    <dependency>
        <groupId>redis.clients</groupId>
        <artifactId>jedis</artifactId>
        <version>2.9.0</version>
    </dependency>
    <dependency>
        <groupId>org.apache.spark</groupId>
        <artifactId>spark-streaming-kafka-0-10_2.11</artifactId>
        <version>2.1.1</version>
    </dependency>
    <dependency>
        <groupId>org.apache.kafka</groupId>
        <artifactId>kafka-clients</artifactId>
        <version>0.10.0.1</version>
    </dependency>
    <dependency>
        <groupId>junit</groupId>
        <artifactId>junit</artifactId>
        <scope>test</scope>
    </dependency>
</dependencies>
```

本章项目数据主要从 Kafka 读取，因此首先搭建 Kafka 集群，而 Kafka 集群需要安装 Zookeeper 集群，以下是详细服务器规划和安装过程。

Kafka 集群部署的核心是配置 config/server.properties 文件，代码如下：

```
broker.id=101  #唯一地标识 int 类型
listeners=PLAINTEXT://server101:9092 #指定本机的地址
log.dirs=/home/isoft/logs/kafka  #指定一个空的已经存在的目录
#指定外部 ZooKeeper 的地址，可选地指定一个子目录只保存 Kafka 的信息
zookeeper.connection=server101,server102,server103:2181/kafka
```

Kafka 和 Zookeeper 集群部署规划如表 10-1 所示。

表10-1 Kafka和Zookeeper集群部署规划

主机/IP/虚拟机名	软 件	进 程	标 识
server101 192.168.56.101 CentOS7-101	zookeeper-3.5.5 kafka_2.11_2.3.1 jdk-1.8	QuorumPeerMan Kafka	myid=101 broker.id=101
server102 192.168.56.102 CentOS7-102	同上	QuorumPeerMan Kafka	myid=102 broker.id=102

(续表)

主机/IP/虚拟机名	软　件	进　程	标　识
server103 192.168.56.103 CentOS7-103	同上	QuorumPeerMan Kafka	myid=103 broker.id=103

操作步骤如下：

步骤01 配置 ZooKeeper 集群并启动。

（1）解压 ZooKeeper：

```
$ tar -zxvf ~/apache-zookeeper-3.5.5-bin.tar.gz -C /app/
```

（2）修改 ZooKeeper 的配置文件<zookeeper_home>/conf/zoo.cfg：

```
tickTime=2000
initLimit=10
syncLimit=5
dataDir=/home/isoft/datas/zk
clientPort=2181
admin.serverPort=9999
server.101=server101:2888:3888
server.102=server102:2888:3888
server.103=server103:2888:3888
```

使用 scp 将文件发送到其他两台机器上

```
$ scp -r zookeeper-3.5.5  server102:/app/
$ scp -r zookeeper-3.5.5  server103:/app/
```

（3）分别在每台服务器的 home/isoft/datas/zk 目录下添加 myid，内容为当前 ZooKeeper 的 id：

```
[hadoop@server101 ~]$ echo 101 > /home/isoft/datas/zk/myid
[hadoop@server102 ~]$ echo 102 > /home/isoft/datas/zk/myid
[hadoop@server103 ~]$ echo 103 > /home/isoft/datas/zk/myid
```

（4）可以使用脚本一次性启动所有 ZooKeeper 集群：

```
#!/bin/bash
if [ $# -lt 1 ]; then
    echo "用法: $0 start | stop | status"
    exit 1
fi
hosts=(server101 server102 server103)
cmd=$1
for host in ${hosts[@]};do
    script="ssh ${host} zkServer.sh ${cmd}"
    echo $script
    eval $script
done
exit 0
```

步骤 02 安装 Kafka。

（1）解压 Kafka：

```
$ tar -zxvf /home/hadoop/kafka_2.12.0-2.7.0.tgz -C /app/
```

（2）修改目录名称：

```
$ mv kafka_2.20.0-2.7.0 kafka-2.7.0
```

（3）修改 Kafka 配置文件 server.properties：

```
broker.id=101    #根据主机的不同，分别设置 broker.id=102,broker.id=103
#根据主机不同，分别设置 server102:9092,server103:9092
listeners=PLAINTEXT://server101:9092
log.dirs=/home/hadoop/logs/kafka
#可以声明一个子目录，便于管理
zookeeper.connect=server101,server102,server103:2181/kafka
```

（4）将 Kafka 分发到其他主机的相同目录下并按上一步的说明进行修改：

```
$ scp -r kafka-2.7.0 server102:/app/
$ scp -r kafka-2.7.0 server103:/app/
```

（5）配置所有主机的环境变量：

```
export KAFKA_HOME=/app/kafka-2.7.0
export PATH=$PATH:$KAFKA_HOME/bin
```

步骤 03 启动 Kafka。

（1）逐台主机上启动 Kafka 服务器：

```
$ kafka-server-start.sh  /app/kafka-2.7.0/config/server.properties  -daemon
```

（2）使用脚本一次性启动：

```
#!/bin/bash
if [ $# -lt 1 ]; then
    echo "使用方法：$0 start | stop"
    exit 1
fi
cmd=$1
servers=(server101 server102 server103)
if [ $cmd == 'start' ]; then
    echo "启动"
    for host in ${servers[@]};
    do
        script="ssh $host kafka-server-start.sh -daemon /app/kafka-2.7.0/config/server.properties"
        echo $script
        eval $script
    done
    exit 0
elif [ $cmd == 'stop' ]; then
```

```
            echo "停止"
            for host in ${servers[@]};do
                script="ssh $host kafka-server-stop.sh
/app/kafka-2.7.0/config/server.properties"
                echo $script
                eval $script
            done
            exit 0
    else
        echo "错误的参数"
        exit 1
    fi
```

(3) Kafka 服务器都启动以后,查看一下所有 broker 的信息。版本信息如下:

```
[hadoop@server101 ~]$ kafka-broker-api-versions.sh --version
2.7.0 (Commit:18a913733fb71c01)
```

输入所有的服务器地址和输入一个服务器地址返回的信息都是一样的:

```
$ kafka-broker-api-versions.sh \
--bootstrap-server server101:9092,server102:9092,server103:9092
```

以下是输入一个服务器地址:

```
[hadoop@server101 ~]$ kafka-broker-api-versions.sh --bootstrap-server
server101:9092
```

返回的信息如下:

```
server103:9092 (id: 103 rack: null) -> (
    Produce(0): 0 to 7 [usable: 7],...
server101:9092 (id: 101 rack: null) -> (
    Produce(0): 0 to 7 [usable: 7],...
server102:9092 (id: 102 rack: null) -> (
    Produce(0): 0 to 7 [usable: 7],...
```

步骤 04 创建并查看 topic。

(1) 创建 topic:

```
$ kafka-topics.sh --create --topic two \
--bootstrap-server server101:9092,server102:9092,server103:9092 \
--partitions 3 --replication-factor 3
```

(2) 查看 topic:

```
$ kafka-topics.sh --describe --topic two --bootstrap-server server101:9092
Topic:two    PartitionCount:3    ReplicationFactor:3
Configs:segment.bytes=1073741824
    Topic: two  Partition: 0    Leader: 101 Replicas: 101,103,102   Isr:
101,103,102
    Topic: two  Partition: 1    Leader: 103 Replicas: 103,102,101   Isr:
103,102,101
    Topic: two  Partition: 2    Leader: 102 Replicas: 102,101,103   Isr:
```

```
102,101,103
```

步骤 05 发布订阅。

（1）创建发布者：

```
$ kafka-console-producer.sh --topic two \
--broker-list server101:9092,server102:9092,server103:9092
>Jack
>mary
>Rose
```

（2）创建消费者：

```
$ kafka-console-consumer.sh --topic two \
--bootstrap-server server101:9092,server102:9092,server103:9092
--from-beginning
Rose
mary
Jack
```

10.2　数据生成模块

首先，我们说明一下本项目的数据来源。本项目使用代码的方式持续地生成模拟数据，然后将数据写入 Kafka 中。Spark Streaming 负责从 Kafka 消费数据，并根据需求对数据进行分析。

模拟出来的数据格式如下：

时间戳,地区,城市,用户 id,广告 id
1566035129449,华南,深圳,101,2

操作步骤如下：

步骤 01 开启 Kafka 集群。

启动 ZooKeeper 和 Kafka 集群，先启动 Zookeeper，再启动 Kafka。

步骤 02 创建 Topic。

在 Kafka 中创建 topic: ads_log。

```
$ kafka-topics.sh --create --topic ads_log\
--bootstrap-server server101:9092,server102:9092,server103:9092 \
--partitions 3 --replication-factor 3
```

步骤 03 创建 spark-realtime 模块，产生循环不断的数据到指定的 topic。

首先导入依赖：

```xml
<dependency>
    <groupId>org.apache.kafka</groupId>
    <artifactId>kafka-clients</artifactId>
```

```
            <version>0.11.0.0</version>
    </dependency>
```

下面开始生成数据。

1）工具类：RandomNumUtil

用于生成随机数。

代码 10-1　RandomNumUtil.scala

```
package com.bigdata.realtime.util
import java.util.Random
import scala.collection.mutable
// 随机生成整数的工具类
object RandomNumUtil {
    val random = new Random()
    //返回一个随机的整数 [from, to]
    def randomInt(from: Int, to: Int): Int = {
        if (from > to) throw new IllegalArgumentException(s"from = $from 应该小于 to = $to")
        // [0, to - from) + from [form, to -from + from ]
        random.nextInt(to - from + 1) + from
    }
    //随机的 Long [from, to]
    def randomLong(from: Long, to: Long): Long = {
        if (from > to) throw new IllegalArgumentException(s"from = $from 应该小于 to = $to")
        random.nextLong().abs % (to - from + 1) + from
    }
    //生成一系列的随机值
    //doop 参数 canReat 表示是否允许随机数重复
    def randomMultiInt(from: Int, to: Int, count: Int, canReat: Boolean = true): List[Int] = {
        if (canReat) {
            (1 to count).map(_ => randomInt(from, to)).toList
        } else {
            val set: mutable.Set[Int] = mutable.Set[Int]()
            while (set.size < count) {
                set += randomInt(from, to)
            }
            set.toList
        }
    }
    def main(args: Array[String]): Unit = {
        println(randomMultiInt(1, 15, 10))
        println(randomMultiInt(1, 8, 10, false))
    }
}
```

2）工具类：RandomOptions

用于生成带有比重的随机选项。

代码 10-2　RandomOptions.scala

```scala
package com.bigdata.realtime.util
import scala.collection.mutable.ListBuffer
//根据提供的值和比重来创建 RandomOptions 对象。
//然后可以通过 getRandomOption 来获取一个随机的预定义的值
object RandomOptions {
    def apply[T](opts: (T, Int)*): RandomOptions[T] = {
        val randomOptions = new RandomOptions[T]()
        randomOptions.totalWeight = (0 /: opts) (_ + _._2) // 计算出总的比重
        opts.foreach {
            case (value, weight) => randomOptions.options ++= (1 to weight).map(_ => value)
        }
        randomOptions
    }
    def main(args: Array[String]): Unit = {
        // 测试
        val opts = RandomOptions(("张三", 10), ("李四", 30), ("ww", 20))

        println(opts.getRandomOption())
        println(opts.getRandomOption())
    }
}
// 工程师 10  程序员 10  老师 20
class RandomOptions[T] {
    var totalWeight: Int = _
    var options = ListBuffer[T]()
    //获取随机的 Option 的值
    def getRandomOption() = {
        options(RandomNumUtil.randomInt(0, totalWeight - 1))
    }
}
```

3）样例类：CityInfo

用于生成样例。

代码 10-3　CityInfo.scala

```scala
package com.bigdata.realtime
//城市表
//city_id:城市 id
//city_name:城市名
//area:城市区域
case class CityInfo(city_id: Long,
                    city_name: String,
                    area: String)
```

4）生成模拟数据：MockRealTime

用于生成模拟数据。

代码 10-4 MockRealtime.scala

```scala
package com.bigdata.realtime.mock
import java.util.Properties
import com.bigdata.realtime.CityInfo
import com.bigdata.realtime.util.{RandomNumUtil, RandomOptions}
import org.apache.kafka.clients.producer.{KafkaProducer, ProducerRecord}
import scala.collection.mutable.ArrayBuffer
//生成实时的模拟数据
object MockRealtime {
    // 数据格式：   timestamp   area    city     userid    adid
    //            某个时间点   某个地区  某个城市  某个用户   某个广告
    def mockRealTimeData(): ArrayBuffer[String] = {
        // 存储模拟的实时数据
        val array = ArrayBuffer[String]()
        // 城市信息
        val randomOpts = RandomOptions(
            (CityInfo(1, "北京", "华北"), 30),
            (CityInfo(2, "上海", "华东"), 30),
            (CityInfo(3, "广州", "华南"), 10),
            (CityInfo(4, "深圳", "华南"), 20),
            (CityInfo(4, "杭州", "华中"), 10))
        (1 to 50).foreach {
            i => {
                val timestamp = System.currentTimeMillis()
                val cityInfo = randomOpts.getRandomOption()
                val area = cityInfo.area
                val city = cityInfo.city_name
                val userid = RandomNumUtil.randomInt(100, 105)
                val adid = RandomNumUtil.randomInt(1, 5)
                array += s"$timestamp,$area,$city,$userid,$adid"
                Thread.sleep(10)
            }
        }
        array
    }
    def createKafkaProducer: KafkaProducer[String, String] = {
        val props: Properties = new Properties
        // Kafka 服务端的主机名和端口号
        props.put("bootstrap.servers",
"server101:9092,server102:9092,server103:9092")
        // key 序列化
        props.put("key.serializer",
"org.apache.kafka.common.serialization.StringSerializer")
        // value 序列化
        props.put("value.serializer",
"org.apache.kafka.common.serialization.StringSerializer")
        new KafkaProducer[String, String](props)
    }
    def main(args: Array[String]): Unit = {
        val topic = "ads_log"
```

```
        val producer: KafkaProducer[String, String] = createKafkaProducer
        while (true) {
            mockRealTimeData().foreach {
                msg => {
                    producer.send(new ProducerRecord(topic, msg))
                    Thread.sleep(100)
                }
            }
            Thread.sleep(1000)
        }
    }
}
```

我们将通过下一节从 Kafka 读取数据来确认 Kafka 中数据是否生成成功。

10.3 从 Kafka 读取数据

编写 RealTimeApp 程序，实现从 Kafka 读取数据的功能。

10.3.1 bean 类 AdsInfo

AdsInfo 类用来封装从 Kafka 读取到的广告点击信息。

代码 10-5 AdsInfo.scala

```
package com.bigdata.realtime.bean
import java.sql.Timestamp
import java.text.SimpleDateFormat
import java.util.Date
case class AdsInfo(ts: Long,
                   area: String,
                   city: String,
                   userId: String,
                   adsId: String,
                   var timestamp: Timestamp = null,
                   var dayString: String = null,  // 2019-12-18
                   var hmString: String = null) { // 11:20
    timestamp = new Timestamp(ts)
    val date = new Date(ts)
    dayString = new SimpleDateFormat("yyyy-MM-dd").format(date)
    hmString = new SimpleDateFormat("HH:mm").format(date)
}
```

10.3.2 工具类 MyKafkaUtil

MyKafkaUtil 类用于获取 Kafka 连接。

代码 10-6　MyKafkaUtil.scala

```scala
import org.apache.kafka.clients.consumer.ConsumerRecord
import org.apache.kafka.common.serialization.StringDeserializer
import org.apache.spark.streaming.StreamingContext
import org.apache.spark.streaming.dstream.InputDStream
import org.apache.spark.streaming.kafka010.{ConsumerStrategies, KafkaUtils, LocationStrategies}
//用于获取 Kafka 连接
object MyKafkaUtil {
    // Kafka 消费者配置
    val kafkaParam = Map(
        "bootstrap.servers" -> "server101:9092,server102:9092,server103:9092",
//用于初始化链接到集群的地址
        "key.deserializer" -> classOf[StringDeserializer],
        "value.deserializer" -> classOf[StringDeserializer],
        //用于标识这个消费者属于哪个消费团体
        "group.id" -> "commerce-consumer-group",
        //如果没有初始化偏移量或者当前的偏移量不存在任何服务器上,那就可以使用这个配置属性
        //使用这个配置,latest 自动重置偏移量为最新的偏移量
        "auto.offset.reset" -> "latest",
        //如果是 true,那么这个消费者的偏移量会在后台自动提交,但是 Kafka 宕机容易丢失数据
        //如果是 false,那么需要手动维护 Kafka 偏移量。本次我们仍然自动维护偏移量
        "enable.auto.commit" -> (true: java.lang.Boolean)
    )
    /*
    创建 DStream,返回接收到的输入数据
    LocationStrategies: 根据给定的主题和集群地址创建 consumer
    LocationStrategies.PreferConsistent: 持续地在所有 Executor 之间分配分区
    ConsumerStrategies: 选择如何在 Driver 和 Executor 上创建和配置 Kafka Consumer
    ConsumerStrategies.Subscribe: 订阅一系列主题
    */
    def getDStream(ssc: StreamingContext, topic: String):
InputDStream[ConsumerRecord[String, String]] = {
        KafkaUtils.createDirectStream[String, String](
            ssc,
            LocationStrategies.PreferConsistent,   // 标配。只要 Kafka 和 Spark 没有部署在一台设备就应该使用这个参数
            ConsumerStrategies.Subscribe[String, String](Array(topic), kafkaParam))
    }
}
```

10.3.3　从 Kafka 消费数据

从 Kafka 消费数据使用 RealTimeApp 类,代码如下:

代码 10-7　RealTimeApp.scala

```scala
import com.bigdata.realtime.app.DayAreaAdsTop
import com.bigdata.realtime.bean.AdsInfo
```

```
import org.apache.kafka.clients.consumer.ConsumerRecord
import org.apache.spark.streaming.StreamingContext
import org.apache.spark.streaming.dstream.{DStream, InputDStream}
import org.apache.spark.{SparkConf, streaming}
import realtime.util.MyKafkaUtil
//从Kafka消费数据
object RealtimeApp {
    def main(args: Array[String]): Unit = {
        val conf: SparkConf = new SparkConf().setMaster("local[*]").setAppName("RealtimeApp")
        val ssc = new StreamingContext(conf, streaming.Seconds(5))
        ssc.checkpoint("realtime")
        val sourceDStream: InputDStream[ConsumerRecord[String, String]] = MyKafkaUtil.getDStream(ssc, "ads_log")
        val adsInfoDSteam: DStream[AdsInfo] = sourceDStream.map(record => {
            // 1576655451922,华北,北京,105,2
            val split: Array[String] = record.value().split(",")
            AdsInfo(
                split(0).toLong,
                split(1),
                split(2),
                split(3),
                split(4))
        })
        // 需求1：每天每地区广告的点击率的Top3
        DayAreaAdsTop.statDayAreaAdsTop3(adsInfoDSteam)
        ssc.start()
        ssc.awaitTermination()
    }
}
```

Kafka 中生成的数据（相当于 Kafka 的消费者来消费数据）如图 10-1 所示。

```
1590328584202,华东,上海,105,4
1590328584224,华东,上海,105,2
1590328584246,华东,上海,101,4
1590328584268,华中,杭州,102,1
1590328584290,华东,上海,103,1
1590328584312,华东,上海,101,2
1590328584333,华东,上海,104,3
1590328584355,华北,北京,100,3
1590328584377,华南,广州,100,5
1590328584399,华东,上海,102,3
...
```

图 10-1

此处是从生成模拟数据 MockRealTime 程序读取的数据，如果读到了数据，可以确认 Kafka 数据生成成功。注意，在测试的时候需要根据自己安装的 Kafka 具体情况来修改 MockRealTime 和 RealtimeApp 类中的 Kafka 的配置信息。

10.4 数据统计实现

最终数据格式使用 hash 存储在 Redis 数据库中。数据统计包括两部分：每天每地区热门广告点击率 Top3 和最近 1 小时内广告点击量实时统计。

10.4.1 每天每地区热门广告点击率 Top3

计算每天每地区热门广告点击率 Top3 实现代码如下：

代码 10-8　AreaAdsClickTop3App.scala

```scala
import com.bigdata.realtime.bean.AdsInfo
import com.bigdata.realtime.util.RedisUtil
import org.apache.spark.streaming.dstream.DStream
import org.json4s.jackson.JsonMethods
import redis.clients.jedis.Jedis
//计算每天每地区热门广告点击率 Top3
object AreaAdsClickTop3App {
  def statAreaClickTop3(adsInfoDStream: DStream[AdsInfo]) = {
    // 1. 每天每地区广告的点击率
    val dayAreaCount: DStream[((String, String), (String, Int))] = adsInfoDStream.map(adsInfo => ((adsInfo.dayString, adsInfo.area, adsInfo.adsId), 1)) // ((天，地区，广告), 1)
      .updateStateByKey((seq: Seq[Int], option: Option[Int]) => Some(seq.sum + option.getOrElse(0))) // ((天，地区，广告), 1000)
      .map {
        case ((day, area, adsId), count) =>
          ((day, area), (adsId, count))
      }
    // 2. 按照（天，地区）分组，然后组内排序，取 Top3
    val dayAreaAdsClickTop3: DStream[(String, String, List[(String, Int)])] = dayAreaCount
      .groupByKey
      .map {
        case ((day, area), adsCountIt) =>
          (day, area, adsCountIt.toList.sortBy(-_._2).take(3))
      }
    // 3. 写入 redis
    dayAreaAdsClickTop3.foreachRDD(rdd => {
      // 建立到 redis 的连接
      val jedisClient: Jedis = RedisUtil.getJedisClient
      val arr: Array[(String, String, List[(String, Int)])] = rdd.collect
      // 写到 redis 的结构举例： key-> "area:das:top3:"+2019-09-25   value:
      //                field              value
      //                {东北: {3: 1000, 2:800, 10:500}  }
```

```scala
      arr.foreach{
        case (day, area, adsIdCountList) => {
          import org.json4s.JsonDSL._
          // list 集合转成 JSON 字符串
          val adsCountJsonString = JsonMethods.compact(JsonMethods.render(adsIdCountList))
          jedisClient.hset(s"area:day:top3:$day", area, adsCountJsonString)
        }
      }
      jedisClient.close()
    })
  }
}
```

10.4.2 最近 1 小时内广告点击量实时统计

统计各广告最近 1 小时内每一分钟的点击量，并存储到 Redis 中，存储格式如表 10-2 所示。

表10-2 点击量统计存储格式

	key
last_hour_ads_click	field: 广告 id　value: {小时分钟:点击次数, 小时分钟 2:点击次数:…}
	field: 11　　　　value:("8:30" :122,"8:31":32,"8:32":910,"8:33":21,…)
	field: 22　　　　value:("8:30" :44,"8:31" :12,"8:32":33."8:33":123,…)

具体实现代码如下：

代码 10-9 LastHourAdsClickApp.scala

```scala
import java.text.SimpleDateFormat
import com.bigdata.realtime.bean.AdsInfo
import com.bigdata.realtime.uitl.RedisUtil
import org.apache.spark.streaming.{Minutes, Seconds}
import org.apache.spark.streaming.dstream.DStream
import org.json4s.jackson.JsonMethods
import redis.clients.jedis.Jedis
//统计各广告最近 1 小时内每一分钟的点击量，并存储到 Redis 中
object LastHourAdsClickApp {
  def statLastHourAdsClick(adsInfoDSteam: DStream[AdsInfo]) = {
    // 统计最近一小时的数据(每分钟点击量)，每 5 秒统计一次
    val windowDStream: DStream[AdsInfo] = adsInfoDSteam.window(Minutes(60), Seconds(5))
    val groupAdsCountDStream: DStream[(String, Iterable[(String, Int)])] = windowDStream.map(adsInfo => {
      ((adsInfo.adsId, adsInfo.hmString), 1)
    }).reduceByKey(_ + _).map {
      case ((adsId, hourMinutes), count) => (adsId, (hourMinutes, count))
    }.groupByKey
    val jsonCountDStream: DStream[(String, String)] = groupAdsCountDStream.map {
```

```
            case (adsId, it) => {
                import org.json4s.JsonDSL._
                val hourMinutesJson: String =
JsonMethods.compact(JsonMethods.render(it))
                (adsId, hourMinutesJson)
            }
        }
        jsonCountDStream.foreachRDD(rdd => {
            val result: Array[(String, String)] = rdd.collect
            import collection.JavaConversions._
            val client: Jedis = RedisUtil.getJedisClient
            client.hmset("last_hour_ads_click", result.toMap)
            client.close()
        })
    }
}
```

根据项目各模块依赖的库，项目总体 pom.xml 代码如下：

代码 10-10　pom.xml

```xml
<dependency>
    <groupId>org.apache.spark</groupId>
    <artifactId>spark-core_2.12</artifactId>
    <version>3.0.0</version>
</dependency>
<dependency>
    <groupId>org.apache.spark</groupId>
    <artifactId>spark-sql_2.12</artifactId>
    <version>3.0.0</version>
</dependency>
<dependency>
    <groupId>org.apache.spark</groupId>
    <artifactId>spark-streaming_2.12</artifactId>
    <version>3.0.0</version>
</dependency>
<dependency>
    <groupId>redis.clients</groupId>
    <artifactId>jedis</artifactId>
    <version>2.9.0</version>
</dependency>
<dependency>
    <groupId>org.apache.spark</groupId>
    <artifactId>spark-streaming-kafka-0-8_2.11</artifactId>
    <version>2.1.1</version>
</dependency>
<dependency>
    <groupId>org.apache.kafka</groupId>
    <artifactId>kafka-clients</artifactId>
    <version>0.11.0.0</version>
</dependency>
```

以上两个功能通过 jedisClient 完成 Redis 数据库的读写。jedisClient 的获取通过 RedisUtil 类获取，RedisUtil 代码如下：

```scala
package com.bigdata.realtime.util
import redis.clients.jedis.{Jedis, JedisPool, JedisPoolConfig}
object RedisUtil{
  var jedisPool:JedisPool=null
  def getJedisClient: Jedis = {
    if(jedisPool==null){
      //println("开辟一个连接池")
      val config = PropertiesUtil.load("config.properties")
      val host = config.getProperty("redis.host")
      val port = config.getProperty("redis.port")

      val jedisPoolConfig = new JedisPoolConfig()
      jedisPoolConfig.setMaxTotal(100)   //最大连接数
      jedisPoolConfig.setMaxIdle(20)     //最大空闲
      jedisPoolConfig.setMinIdle(20)     //最小空闲
      jedisPoolConfig.setBlockWhenExhausted(true)  //忙碌时是否等待
      jedisPoolConfig.setMaxWaitMillis(500)//忙碌时等待时长（毫秒）
      jedisPoolConfig.setTestOnBorrow(true)  //对每次获得的连接进行测试

      jedisPool=new JedisPool(jedisPoolConfig,host,port.toInt)
    }
    //println(s"jedisPool.getNumActive = ${jedisPool.getNumActive}")
    //println("获得一个连接")
    jedisPool.getResource
  }
}
```

第 11 章

电影影评大数据分析项目实战

本实战项目基于 Spark SQL 对电影数据和电影评分数据进行数据分析，首先获取数据源，转换成 DataFrame，并调用封装好的业务逻辑类来生成临时视图，然后基于 Spark SQL 结合不同的业务需求编写不同的 SQL 语句，完成逻辑代码，实现电影类别及评分等数据统计。

本章主要知识点：

- 利用 Spark SQL 进行数据分析
- 统计结果写入 MySQL
- 读取 CSV 文件到 DataSet

11.1 项目介绍

1. 数据集介绍

使用 MovieLens 的名称为 ml-25m.zip 的数据集，使用的文件是 movies.csv 和 ratings.csv，上述文件的下载地址为：

```
http://files.grouplens.org/datasets/movielens/ml-25m.zip
```

- movies.csv

该文件是电影数据，对应的是维度并能表数据，大小为 2.89MB，包括 6 万多部电影，其数据格式为[movieId,title,genres]，分别对应**[电影 id, 电影名称, 电影所属分类]**，样例数据如下（分隔符为逗号）：

```
1,Toy Story (1995),Adventure|Animation|Children|Comedy|Fantasy
```

● ratings.csv

该文件为电影评分数据，对应的是事实表数据，大小为 646MB，其数据格式为 [userId,movieId,rating,timestamp]，分别对应[用户 id,电影 id,评分,时间戳]，样例数据如下（分隔符为英文逗号）：

```
1,296,5,1147880044
```

项目结构如图 11-1 所示。

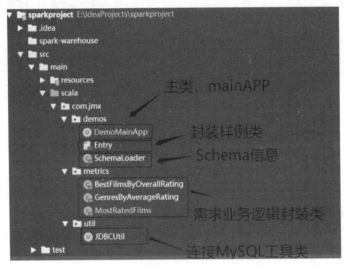

图 11-1

2. 项目需求

本影评分析项目需要实现以下 3 个需求：

- 需求 1：查找电影评分个数超过 5000 且平均评分最高的前 10 部电影的名称及其对应的平均评分。
- 需求 2：查找每个电影类别及其对应的平均评分。
- 需求 3：查找评分次数最多的前 10 部电影。

11.2 项目实现

本章项目使用本地模式，不需要安装 Spark，引入相关 JAR 包即可。本章需要引入的所有包的依赖参见 pom.xml，代码如下：

```
<dependencies>
    <dependency>
        <groupId>org.scala-lang</groupId>
        <artifactId>scala-library</artifactId>
        <version>${scala.version}</version>
```

```xml
    </dependency>
    <dependency>
        <groupId>c3p0</groupId>
        <artifactId>c3p0</artifactId>
        <version>0.9.1.2</version>
    </dependency>
    <dependency>
        <groupId>commons-dbutils</groupId>
        <artifactId>commons-dbutils</artifactId>
        <version>1.6</version>
    </dependency>
    <dependency>
        <groupId>junit</groupId>
        <artifactId>junit</artifactId>
        <version>4.12</version>
        <scope>test</scope>
    </dependency>
    <dependency>
        <groupId>org.specs</groupId>
        <artifactId>specs</artifactId>
        <version>1.2.5</version>
        <scope>test</scope>
    </dependency>
    <!-- https://mvnrepository.com/artifact/org.apache.spark/spark-core -->
    <dependency>
        <groupId>org.apache.spark</groupId>
        <artifactId>spark-core_2.11</artifactId>
        <version>2.4.3</version>
    </dependency>
    <!-- https://mvnrepository.com/artifact/org.apache.spark/spark-sql -->
    <dependency>
        <groupId>org.apache.spark</groupId>
        <artifactId>spark-sql_2.11</artifactId>
        <version>2.4.3</version>
    </dependency>
    <!-- https://mvnrepository.com/artifact/org.apache.spark/spark-streaming -->
    <dependency>
        <groupId>org.apache.spark</groupId>
        <artifactId>spark-streaming_2.11</artifactId>
        <version>2.4.3</version>
    </dependency>
    <!-- https://mvnrepository.com/artifact/org.apache.spark/spark-mllib -->
    <dependency>
        <groupId>org.apache.spark</groupId>
        <artifactId>spark-mllib_2.11</artifactId>
        <version>2.4.3</version>
        <!--<scope>runtime</scope>-->
    </dependency>
    <!-- https://mvnrepository.com/artifact/org.apache.spark/spark-streaming-kafka-0-10 -->
```

```xml
<dependency>
    <groupId>org.apache.spark</groupId>
    <artifactId>spark-streaming-kafka-0-10_2.11</artifactId>
    <version>2.4.3</version>
</dependency>
<!-- https://mvnrepository.com/artifact/org.apache.spark/spark-hive -->
<dependency>
    <groupId>org.apache.spark</groupId>
    <artifactId>spark-hive_2.11</artifactId>
    <version>2.4.3</version>
</dependency>

<!-- https://mvnrepository.com/artifact/mysql/mysql-connector-java -->
<dependency>
    <groupId>mysql</groupId>
    <artifactId>mysql-connector-java</artifactId>
    <version>5.1.39</version>
</dependency>
<!-- https://mvnrepository.com/artifact/org.apache.hadoop/hadoop-common -->
<dependency>
    <groupId>org.apache.hadoop</groupId>
    <artifactId>hadoop-common</artifactId>
    <version>2.7.7</version>
</dependency>
<!-- https://mvnrepository.com/artifact/org.apache.hadoop/hadoop-client -->
<dependency>
    <groupId>org.apache.hadoop</groupId>
    <artifactId>hadoop-client</artifactId>
    <version>2.7.7</version>
</dependency>
<!-- https://mvnrepository.com/artifact/org.apache.hadoop/hadoop-hdfs -->
<dependency>
    <groupId>org.apache.hadoop</groupId>
    <artifactId>hadoop-hdfs</artifactId>
    <version>2.7.7</version>
</dependency>
<dependency>
    <groupId>org.apache.avro</groupId>
    <artifactId>avro-tools</artifactId>
    <version>1.8.1</version>
</dependency>
<!-- https://mvnrepository.com/artifact/org.apache.hive/hive-cli -->
<dependency>
    <groupId>org.apache.hive</groupId>
    <artifactId>hive-cli</artifactId>
    <version>2.3.4</version>
</dependency>
<dependency>
```

```xml
        <groupId>org.apache.hive</groupId>
        <artifactId>hive-exec</artifactId>
        <version>2.3.4</version>
    </dependency>
    <dependency>
        <groupId>org.apache.commons</groupId>
        <artifactId>commons-dbcp2</artifactId>
        <version>2.1.1</version>
    </dependency>
    <dependency>
        <groupId>redis.clients</groupId>
        <artifactId>jedis</artifactId>
        <version>2.8.0</version>
    </dependency>
    <dependency>
        <groupId>ru.yandex.clickhouse</groupId>
        <artifactId>clickhouse-jdbc</artifactId>
        <version>0.2.4</version>
    </dependency>
    <!-- https://mvnrepository.com/artifact/com.google.guava/guava -->
    <dependency>
        <groupId>com.google.guava</groupId>
        <artifactId>guava</artifactId>
        <version>28.0-jre</version>
    </dependency>
</dependencies>
```

以上依赖是基于 Scala 2.13 版本，所有的 Spark 依赖包均与 Scala 2.13 一致，本章提供的项目源码可以在 Windows 本地导入到 IDEA 后直接运行。

11.2.1 公共代码开发

1）DemoMainApp

主程序 DemoMainApp 是程序执行的入口，主要用于获取数据源，转换成 DataFrame，并调用封装好的业务逻辑类。代码如下：

代码 11-1 DemoMainApp.scala

```scala
object DemoMainApp {
  // 文件路径
  private val MOVIES_CSV_FILE_PATH = "movies.csv"
  private val RATINGS_CSV_FILE_PATH = "ratings.csv"
  def main(args: Array[String]): Unit = {
    // 创建 Spark Session
    val spark = SparkSession
      .builder
      .master("local[4]")
      .getOrCreate
    // schema 信息
    val schemaLoader = new SchemaLoader
```

```
    // 读取 Movie 数据集
    val movieDF = readCsvIntoDataSet(spark, MOVIES_CSV_FILE_PATH,
schemaLoader.getMovieSchema)
    // 读取 Rating 数据集
    val ratingDF = readCsvIntoDataSet(spark, RATINGS_CSV_FILE_PATH,
schemaLoader.getRatingSchema)
    // 需求 1：查找电影评分个数超过 5000 且平均评分最高的前 10 部电影名称及其对应的平均评分
    val bestFilmsByOverallRating = new BestFilmsByOverallRating
    //bestFilmsByOverallRating.run(movieDF, ratingDF, spark)
    // 需求 2：查找每个电影类别及其对应的平均评分
    val genresByAverageRating = new GenresByAverageRating
    //genresByAverageRating.run(movieDF, ratingDF, spark)
    // 需求 3：查找评分次数最多的前 10 部电影
    val mostRatedFilms = new MostRatedFilms
    mostRatedFilms.run(movieDF, ratingDF, spark)
    spark.close()
  }
  //读取数据文件，转换成 DataFrame
  //参数为：spark,path,schema
  def readCsvIntoDataSet(spark: SparkSession, path: String, schema: StructType)
= {
    val dataSet = spark.read
      .format("csv")
      .option("header", "true")
      .schema(schema)
      .load(path)
    dataSet
  }
}
```

2）Entry

Entry 类为实体类，封装了数据源的样例类和结果表的样例类，代码如下：

代码 11-2　Entry.scala

```
class Entry {
}
case class Movies(
              movieId: String,    // 电影的 id
              title: String,      // 电影的标题
              genres: String      // 电影类别
            )
case class Ratings(
              userId: String,     // 用户的 id
              movieId: String,    // 电影的 id
              rating: String,     // 用户评分
              timestamp: String   // 时间戳
            )
// 需求 1 的 MySQL 结果表
case class tenGreatestMoviesByAverageRating(
```

```
                movieId: String,    // 电影的 id
                title: String,      // 电影的标题
                avgRating: String   // 电影平均评分
                )
// 需求 2 的 MySQL 结果表
case class topGenresByAverageRating(
                genres: String,     //电影类别
                avgRating: String   // 平均评分
                )
// 需求 3 的 MySQL 结果表
case class tenMostRatedFilms(
                movieId: String,    // 电影的 id
                title: String,      // 电影的标题
                ratingCnt: String   // 电影被评分的次数
                )
```

3）SchemaLoader

SchemaLoader 类封装了数据集的 schema 信息，主要用于读取数据源时指定 schema 信息，该类代码如下：

代码 11-3　SchemaLoader.scala

```
class SchemaLoader {
  // movies 数据集 schema 信息
  private val movieSchema = new StructType()
    .add("movieId", DataTypes.StringType, false)
    .add("title", DataTypes.StringType, false)
    .add("genres", DataTypes.StringType, false)
  // ratings 数据集 schema 信息
  private val ratingSchema = new StructType()
    .add("userId", DataTypes.StringType, false)
    .add("movieId", DataTypes.StringType, false)
    .add("rating", DataTypes.StringType, false)
    .add("timestamp", DataTypes.StringType, false)
  def getMovieSchema: StructType = movieSchema
  def getRatingSchema: StructType = ratingSchema
}
```

4）JDBCUtil

JDBCUtil 类封装了连接 MySQL 的逻辑，主要用于连接 MySQL，在业务逻辑代码中会使用该工具类获取 MySQL 连接，将结果数据写入 MySQL 中。该类代码如下：

代码 11-4　JDBCUtil.scala

```
object JDBCUtil {
  val dataSource = new ComboPooledDataSource()
  val user = "root"   //MySQL 用户名
  val password = "root"   //MySQL 密码
```

```scala
val url = "jdbc:mysql://localhost:3306/mydb"  //MySQL 数据库 url
dataSource.setUser(user)
dataSource.setPassword(password)
dataSource.setDriverClass("com.mysql.cj.jdbc.Driver")
dataSource.setJdbcUrl(url)
dataSource.setAutoCommitOnClose(false)// 获取连接
def getQueryRunner(): Option[QueryRunner]={
  try {
    Some(new QueryRunner(dataSource))
  }catch {
    case e:Exception =>
      e.printStackTrace()
      None
  }
}
```

11.2.2 平均评分最高的前 10 部电影

BestFilmsByOverallRating.scala 是需求 1 实现的业务逻辑封装，该类中有一个 run()方法，主要用于封装计算逻辑。

代码 11-5　BestFilmsByOverallRating.scala

```scala
//需求 1：查找电影评分个数超过 5000 且平均评分最高的前 10 部电影名称及其对应的平均评分
class BestFilmsByOverallRating extends Serializable {
  def run(moviesDataset: DataFrame, ratingsDataset: DataFrame, spark: SparkSession) = {
    import spark.implicits._
    // 将 moviesDataset 注册成表
    moviesDataset.createOrReplaceTempView("movies")
    // 将 ratingsDataset 注册成表
    ratingsDataset.createOrReplaceTempView("ratings")
    val ressql1 =
      """
        |WITH ratings_filter_cnt AS (
        |SELECT
        |     movieId,
        |     count( * ) AS rating_cnt,
        |     avg( rating ) AS avg_rating
        |FROM
        |     ratings
        |GROUP BY
        |     movieId
        |HAVING
        |     count( * ) >= 5000
        |),
```

```
      |ratings_filter_score AS (
      |SELECT
      |    movieId, -- 电影id
      |    avg_rating -- 电影平均评分
      |FROM ratings_filter_cnt
      |ORDER BY avg_rating DESC -- 平均评分降序排序
      |LIMIT 10 -- 平均分最高的前10部电影
      |)
      |SELECT
      |    m.movieId,
      |    m.title,
      |    r.avg_rating AS avgRating
      |FROM
      |    ratings_filter_score r
      |JOIN movies m ON m.movieId = r.movieId
    """.stripMargin
    val resultDS = spark.sql(ressql1).as[tenGreatestMoviesByAverageRating]
    // 打印数据
    resultDS.show(10)
    resultDS.printSchema()
    // 写入MySQL
    resultDS.foreachPartition(par => par.foreach(insert2Mysql(_)))
  }
  //获取连接,调用写入MySQL数据的方法
  private def insert2Mysql(res: tenGreatestMoviesByAverageRating): Unit = {
    lazy val conn = JDBCUtil.getQueryRunner()
    conn match {
      case Some(connection) => {
        upsert(res, connection)
      }
      case None => {
        println("Mysql连接失败")
        System.exit(-1)
      }
    }
  }
  //封装将结果写入MySQL的方法,执行写入操作
  private def upsert(r: tenGreatestMoviesByAverageRating, conn: QueryRunner): Unit = {
    try {
      val sql =
        s"""
          |REPLACE INTO `ten_movies_averagerating`(
          |movieId,
          |title,
          |avgRating
```

```
        |)
        |VALUES
        |(?,?,?)
      """.stripMargin
    // 执行 insert 操作
    conn.update(
      sql,
      r.movieId,
      r.title,
      r.avgRating
    )
  } catch {
    case e: Exception => {
      e.printStackTrace()
      System.exit(-1)
    }
  }
 }
}
```

需求 1 结果表建表语句如下:

```
CREATE TABLE `ten_movies_averagerating` (
  `id` int(11) NOT NULL AUTO_INCREMENT COMMENT '自增 id',
  `movieId` int(11) NOT NULL COMMENT '电影 id',
  `title` varchar(100) NOT NULL COMMENT '电影名称',
  `avgRating` decimal(10,2) NOT NULL COMMENT '平均评分',
  `update_time` datetime DEFAULT CURRENT_TIMESTAMP COMMENT '更新时间',
  PRIMARY KEY (`id`),
  UNIQUE KEY `movie_id_UNIQUE` (`movieId`)
) ENGINE=InnoDB DEFAULT CHARSET=utf8;
```

统计结果，平均评分最高的前 10 部电影如表 11-1 所示。

表11-1　平均评分最高的前10部电影

movieId	title	avgRating
318	The Shawshank Redemption (1994)	4.41
858	The Godfather (1972)	4.32
50	The Usual Suspects (1995)	4.28
1221	The Godfather: Part II (1974)	4.26
527	Schindler's List (1993)	4.25
2019	Seven Samurai (Shichinin no samurai) (1954)	4.25
904	Rear Window (1954)	4.24
1203	12 Angry Men (1957)	4.24
2959	Fight Club (1999)	4.23
1193	One Flew Over the Cuckoo's Nest (1975)	4.22

上述电影评分对应的电影中文名称如表 11-2 所示。

表11-2 平均评分最高的前10部电影的英文名称及对应的中文名称

英 文 名 称	中 文 名 称
The Shawshank Redemption (1994)	肖申克的救赎
The Godfather (1972)	教父1
The Usual Suspects (1995)	非常嫌疑犯
The Godfather: Part II (1974)	教父2
Schindler's List (1993)	辛德勒的名单
Seven Samurai (Shichinin no samurai) (1954)	七武士
Rear Window (1954)	后窗
12 Angry Men (1957)	十二怒汉
Fight Club (1999)	搏击俱乐部
One Flew Over the Cuckoo's Nest (1975)	飞越疯人院

11.2.3 电影类别及其平均评分

GenresByAverageRating.scala 是需求2实现的业务逻辑封装，该类中有一个 run() 方法，主要用于封装计算逻辑。

代码 11-6　GenresByAverageRating.scala

```
//需求2：查找每个电影类别及其对应的平均评分
class GenresByAverageRating extends Serializable {
  def run(moviesDataset: DataFrame, ratingsDataset: DataFrame, spark: SparkSession) = {
    import spark.implicits._
    // 将 moviesDataset 注册成表
    moviesDataset.createOrReplaceTempView("movies")
    // 将 ratingsDataset 注册成表
    ratingsDataset.createOrReplaceTempView("ratings")
    val ressql2 =
      """
        |WITH explode_movies AS (
        |SELECT
        |movieId,
        |title,
        |category
        |FROM
        |movies lateral VIEW explode ( split ( genres, "\\|" ) ) temp AS category
        |)
        |SELECT
        |m.category AS genres,
        |avg( r.rating ) AS avgRating
        |FROM
        |explode_movies m
        |JOIN ratings r ON m.movieId = r.movieId
        |GROUP BY
        |m.category
        | """.stripMargin
    val resultDS = spark.sql(ressql2).as[topGenresByAverageRating]
```

```scala
    // 打印数据
    resultDS.show(10)
    resultDS.printSchema()
    // 写入 MySQL
    resultDS.foreachPartition(par => par.foreach(insert2Mysql(_)))
  }
  //获取连接，调用写入 MySQL 数据的方法
  private def insert2Mysql(res: topGenresByAverageRating): Unit = {
    lazy val conn = JDBCUtil.getQueryRunner()
    conn match {
      case Some(connection) => {
        upsert(res, connection)
      }
      case None => {
        println("Mysql 连接失败")
        System.exit(-1)
      }
    }
  }
  //封装将结果写入 MySQL 的方法，执行写入操作
  private def upsert(r: topGenresByAverageRating, conn: QueryRunner): Unit = {
    try {
      val sql =
        s"""
           |REPLACE INTO `genres_average_rating`(
           |genres,
           |avgRating
           |)
           |VALUES
           |(?,?)
      """.stripMargin
      // 执行 insert 操作
      conn.update(
        sql,
        r.genres,
        r.avgRating
      )
    } catch {
      case e: Exception => {
        e.printStackTrace()
        System.exit(-1)
      }
    }
  }
}
```

需求 2 结果表建表语句如下：

```sql
CREATE TABLE genres_average_rating (
    `id` INT ( 11 ) NOT NULL AUTO_INCREMENT COMMENT '自增id',
    `genres` VARCHAR ( 100 ) NOT NULL COMMENT '电影类别',
    `avgRating` DECIMAL ( 10, 2 ) NOT NULL COMMENT '电影类别平均评分',
    `update_time` datetime DEFAULT CURRENT_TIMESTAMP COMMENT '更新时间',PRIMARY
```

```
  KEY ( `id` ),
    UNIQUE KEY `genres_UNIQUE` ( `genres` )
  ) ENGINE = INNODB DEFAULT CHARSET = utf8;
```

统计结果中共有 20 个电影分类，每个电影分类的平均评分如表 11-3 所示。

表11-3 电影类别及其平均评分

genres	avgRating
Film-Noir	3.93
War	3.79
Documentary	3.71
Crime	3.69
Drama	3.68
Mystery	3.67
Animation	3.61
IMAX	3.6
Western	3.59
Musical	3.55
Romance	3.54
Adventure	3.52
Thriller	3.52
Fantasy	3.51
Sci-Fi	3.48
Action	3.47
Children	3.43
Comedy	3.42
(no genres listed)	3.33
Horror	3.29

电影分类对应的中文名称如表 11-4 所示。

表11-4 电影分类及其中文名称

分 类	中 文 名 称
Film-Noir	黑色电影
War	战争
Documentary	纪录片
Crime	犯罪
Drama	历史剧
Mystery	推理
Animation	动画片
IMAX	巨幕电影
Western	西部电影
Musical	音乐
Romance	浪漫
Adventure	冒险

（续表）

分　　类	中 文 名 称
Thriller	惊悚片
Fantasy	魔幻电影
Sci-Fi	科幻
Action	动作
Children	儿童
Comedy	喜剧
(no genres listed)	未分类
Horror	恐怖

11.2.4　评分次数最多的前 10 部电影

MostRatedFilms.scala 是需求 3 实现的业务逻辑封装，该类中有一个 run()方法，主要用于封装计算逻辑。

代码 11-7　MostRatedFilms.scala

```scala
//需求 3：查找评分次数最多的前 10 部电影
class MostRatedFilms extends Serializable {
    def run(moviesDataset: DataFrame, ratingsDataset: DataFrame,spark: SparkSession) = {
        import spark.implicits._
        // 将 moviesDataset 注册成表
        moviesDataset.createOrReplaceTempView("movies")
        // 将 ratingsDataset 注册成表
        ratingsDataset.createOrReplaceTempView("ratings")
    val ressql3 =
      """
        |WITH rating_group AS (
        |   SELECT
        |       movieId,
        |       count( * ) AS ratingCnt
        |   FROM ratings
        |   GROUP BY movieId
        |),
        |rating_filter AS (
        |   SELECT
        |       movieId,
        |       ratingCnt
        |   FROM rating_group
        |   ORDER BY ratingCnt DESC
        |   LIMIT 10
        |)
        |SELECT
        |   m.movieId,
        |   m.title,
        |   r.ratingCnt
        |FROM
```

```scala
      |    rating_filter r
      |JOIN movies m ON r.movieId = m.movieId
      |
    """.stripMargin
    val resultDS = spark.sql(ressql3).as[tenMostRatedFilms]
    // 打印数据
    resultDS.show(10)
    resultDS.printSchema()
    // 写入 MySQL
    resultDS.foreachPartition(par => par.foreach(insert2Mysql(_)))
  }
  //获取连接，调用写入 MySQL 数据的方法
  private def insert2Mysql(res: tenMostRatedFilms): Unit = {
    lazy val conn = JDBCUtil.getQueryRunner()
    conn match {
      case Some(connection) => {
        upsert(res, connection)
      }
      case None => {
        println("Mysql 连接失败")
        System.exit(-1)
      }
    }
  }
  //封装将结果写入 MySQL 的方法，执行写入操作
  private def upsert(r: tenMostRatedFilms, conn: QueryRunner): Unit = {
    try {
      val sql =
        s"""
           |REPLACE INTO `ten_most_rated_films`(
           |movieId,
           |title,
           |ratingCnt
           |)
           |VALUES
           |(?,?,?)
           """.stripMargin
      // 执行 insert 操作
      conn.update(
        sql,
        r.movieId,
        r.title,
        r.ratingCnt
      )
    } catch {
      case e: Exception => {
        e.printStackTrace()
        System.exit(-1)
      }
    }
  }
```

```
    }
  }
```

需求 3 统计结果表的创建语句如下：

```
CREATE TABLE ten_most_rated_films (
    `id` INT ( 11 ) NOT NULL AUTO_INCREMENT COMMENT '自增 id',
    `movieId` INT ( 11 ) NOT NULL COMMENT '电影 Id',
    `title` varchar(100) NOT NULL COMMENT '电影名称',
    `ratingCnt` INT(11) NOT NULL COMMENT '电影被评分的次数',
    `update_time` datetime DEFAULT CURRENT_TIMESTAMP COMMENT '更新时间',PRIMARY KEY ( `id` ),
    UNIQUE KEY `movie_id_UNIQUE` ( `movieId` )
) ENGINE = INNODB DEFAULT CHARSET = utf8;
```

统计结果如表 11-5 所示。

表11-5　评分数最多的前10部电影

movieId	title	ratingCnt
356	Forrest Gump (1994)	81491
318	The Shawshank Redemption (1994)	81482
296	Pulp Fiction (1994)	79672
593	The Silence of the Lambs (1991)	74127
2571	The Matrix (1999)	72674
260	Star Wars: Episode IV - A New Hope (1977)	68717
480	Jurassic Park (1993)	64144
527	Schindler's List (1993)	60411
110	Braveheart (1995)	59184
2959	Fight Club (1999)	58773

评分次数最多的前 10 部电影对应的中文名称如表 11-6 所示。

表11-6　评分次数最多的前10部电影对应的中文名称

英 文 名 称	中 文 名 称
Forrest Gump (1994)	阿甘正传
The Shawshank Redemption (1994)	肖申克的救赎
Pulp Fiction (1994)	低俗小说
The Silence of the Lambs (1991)	沉默的羔羊
The Matrix (1999)	黑客帝国
Star Wars: Episode IV - A New Hope (1977)	星球大战
Jurassic Park (1993)	侏罗纪公园
Schindler's List (1993)	辛德勒的名单
Braveheart (1995)	勇敢的心
Fight Club (1999)	搏击俱乐部